AC Electrical Circuit Analysis

Laboratory Manual

James M. Fiore

Laboratory Manual

for

AC Electrical Circuit Analysis

by

James M. Fiore

Version 2.3.5, 07 March 2021

Published by James M. Fiore via dissidents

ISBN13: 978-1796526639

For more information or feedback, contact:

James Fiore, Professor
Electrical Engineering Technology
Mohawk Valley Community College
1101 Sherman Drive
Utica, NY 13501
jfiore@mvcc.edu

For the latest revisions, related titles, and links to low cost print versions, go to:

www.mvcc.edu/jfiore or www.dissidents.com

YouTube Channel: *Electronics with Professor Fiore*

Cover art, *Chapman's Contribution Redux*, by the author

Introduction

This laboratory manual is intended for use in an AC electrical circuits course and is appropriate for either a two or four year electrical engineering technology curriculum. The manual contains sufficient exercises for a typical 15 week course using a two to three hour practicum period. The topics range from introductory RL and RC circuits and oscilloscope orientation through series-parallel circuits, superposition, Thevenin's theorem, maximum power transfer theorem, and concludes with series and parallel resonance. For equipment, each lab station should include a dual channel oscilloscope (preferably digital), a function generator and a quality DMM. The exercise covering superposition requires two function generators. For components, a selection of standard value ¼ watt carbon film resistors ranging from a few ohms to a few mega ohms is required along with a selection of film capacitors up to 2.2 µF, and 1 mH and 10 mH inductors. A decade resistance box may also be useful.

Each exercise begins with an Objective and a Theory Overview. The Equipment List follows with space provided for serial numbers and measured values of components. Schematics are presented next along with the step-by-step procedure. All data tables are grouped together, typically with columns for the theoretical and experimental results, along with a column for the percent deviations between them. Finally, a group of appropriate questions are presented. For those with longer scheduled lab times, a useful addition is to simulate the circuit(s) with a SPICE-based tool such as Multisim or PSpice, TINA-TI, LTspice, or similar software, and compare those results to the theoretical and experimental results as well.

A companion laboratory manual for DC electrical circuits is also available. Other manuals in this series include Semiconductor Devices (diodes, bipolar transistors and FETs), Operational Amplifiers & Linear Integrated Circuits, Computer Programming with Python™ and Multisim™, and Embedded Controllers Using C and Arduino. Texts are available for DC and AC Electrical Circuit Analysis, Embedded Controllers, Op Amps & Linear Integrated Circuits, and Semiconductor Devices.

A Note from the Author

This work was borne out of the frustration of finding a lab manual that covered all of the appropriate material at sufficient depth while remaining readable and affordable for the students. It is used at Mohawk Valley Community College in Utica, NY, for our ABET accredited AAS program in Electrical Engineering Technology. I am indebted to my students, co-workers and the MVCC family for their support and encouragement of this project. I thank Mr. Bill Hunt in particular for his suggestions that led to revision 1.2. While it would have been possible to seek a traditional publisher for this work, as a long-time supporter and contributor to freeware and shareware computer software, I have decided instead to release this using a Creative Commons non-commercial, share-alike license. I encourage others to make use of this manual for their own work and to build upon it. If you do add to this effort, I would appreciate a notification.

"Violence is the last refuge of the incompetent."

- Isaac Asimov

Table of Contents

1

Introduction to RL and RC Circuits

Objective

In this exercise, the DC steady state response of simple RL and RC circuits is examined. The transient behavior of RC circuits is also tested.

Theory Overview

The DC steady state response of RL and RC circuits are essential opposite of each other: that is, once steady state is reached, capacitors behave as open circuits while inductors behave as short circuits. In practicality, steady state is reached after five time constants. The time constant for an RC circuit is simply the effective capacitance times the effective resistance, $\tau = RC$. In the inductive case, the time constant is the effective inductance divided by the effective resistance, $\tau = L/R$.

Equipment

(1) DC power supply	model:____________	srn:____________
(1) DMM	model:____________	srn:____________
(1) Stop watch		

Components

(1) 1 µF	actual:____________
(1) 470 µF	actual:____________
(1) 10 mH	actual:____________
(1) 10 kΩ	actual:____________
(1) 47 kΩ	actual:____________

Schematics

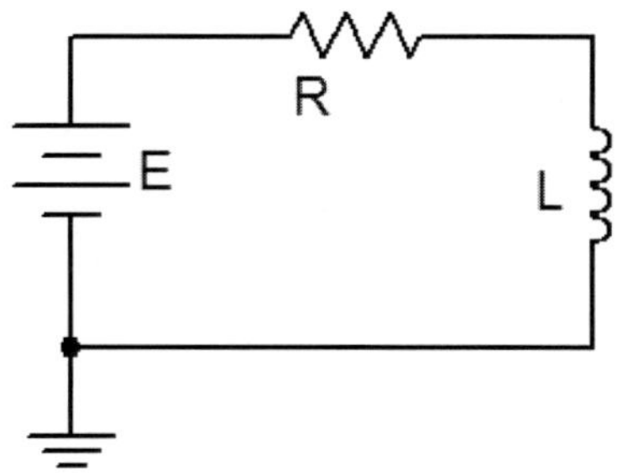

Figure 1.1

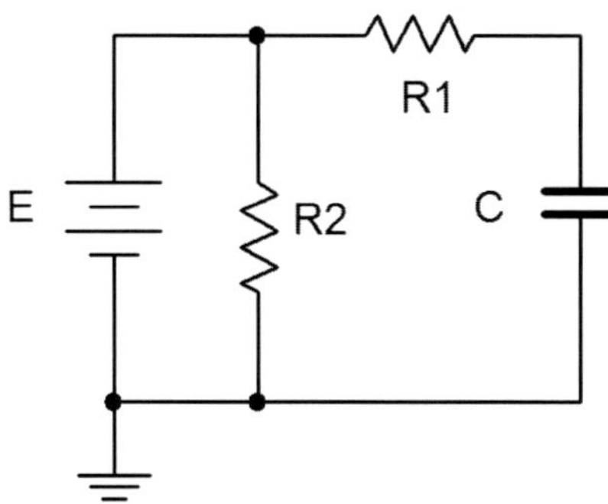

Figure 1.2

Procedure

RL Circuit

1. Using figure 1.1 with E=10 V, R=47 kΩ, and L=10 mH, calculate the time constant and record it in Table 1.1. Also, calculate and record the expected steady state inductor voltage in Table 1.2.

2. Set the power supply to 10 V but do not hook it up to the remainder of the circuit. After connecting the resistor and inductor, connect the DMM across the inductor set to read DC voltage (20 volt scale).

3. Connect the power supply to the circuit. The circuit should reach steady state very quickly, in much less than one second. Record the experimental inductor voltage in Table 1.2. Also, compute and record the percent deviation between experimental and theory in Table 1.2.

RC Circuit

4. Using figure 1.2 with E=10 V, R1=47 kΩ, R2=10k and C=1 μF, calculate the time constant and record it in Table 1.3. Also, calculate and record the expected steady state capacitor voltage in Table 1.4.

5. Set the power supply to 10 V but do not hook it up to the remainder of the circuit. After connecting the resistors and capacitor, connect the DMM across the capacitor set to read DC voltage (20 volt scale).

6. Connect the power supply to the circuit. The circuit should reach steady state quickly, in under one second. Record the experimental capacitor voltage in Table 1.4. Also, compute and record the percent deviation between experimental and theory in Table 1.4.

RC Circuit (long time constant)

7. Using figure 1.2 with E=10 V, R1=47k, R2=10 kΩ and C=470 μF, calculate the time constants and record them in Table 1.5. Also, calculate and record the expected steady state capacitor voltage (charge phase) in Table 1.5.

8. Set the power supply to standby, and after waiting a moment for the capacitor to discharge, remove the capacitor and replace it with the 470 μF. Connect the DMM across the capacitor set to read DC voltage (20 volt scale).

9. Energize the circuit and record the capacitor voltage every 10 seconds as shown in Table 1.6. This is the charge phase.

10. Disconnect the power supply from the circuit and record the capacitor voltage every 10 seconds as shown in Table 1.7. This is the discharge phase.

11. Using the data from Tables 1.6 and 1.7, create two plots of capacitor voltage versus time and compare them to the theoretical plots found in the text.

Data Tables

Table 1.1

V$_L$ Theory	V$_L$ Experimental	Deviation

Table 1.2

Table 1.3

V$_C$ Theory	V$_C$ Experimental	Deviation

Table 1.4

τ charge	
τ discharge	
V$_C$ Theory	

Table 1.5

Time (sec)	Voltage
0	
10	
20	
30	
40	
50	
60	
70	
80	
90	
100	
110	
120	

Table 1.6

Time (sec)	Voltage
0	
10	
20	
30	
40	
50	
60	
70	
80	
90	
100	
110	
120	
130	
140	
150	

Table 1.7

Questions

1. What is a reasonable approximation for an inductor at DC steady state?

2. What is a reasonable approximation for a capacitor at DC steady state?

3. How can a reasonable approximation for time-to-steady state of an RC circuit be computed?

4. In general, what sorts of shapes do the charge and discharge voltages of DC RC circuits follow?

   Laboratory Manual for AC Electrical Circuit Analysis

2

Phasor and Vector Review

Objective

The proper manipulation and representation of vectors is paramount for AC circuit analysis. Addition, subtraction, multiplication and division of vectors in both rectangular and polar forms are examined in both algebraic and graphical forms. Representations of waveforms using both phasor and time domain graphs are also examined.

Procedure

Perform the following operations, including phasor diagrams where appropriate.

1. $(6 + j10) + (8 - j2)$
2. $(2 + j5) - (10 - j4)$
3. $10\angle 0° + 20\angle 90°$
4. $10\angle 45° + 2\angle{-30°}$
5. $20\angle 10° - 5\angle 75°$
6. $(10 + j20) * (5 + j5)$
7. $(2 + j10) / (0.5 + j2)$
8. $10\angle 0° * 10\angle 90°$
9. $10\angle 45° * 10\angle{-45°}$
10. $10\angle 90° / 5\angle 10°$
11. $10\angle 90° / 40\angle{-40°}$
12. $1 / 200\angle 90°$

Draw the following expressions as time domain graphs.

13. $v = 10 \sin 2\pi 100t$
14. $v = 20 \sin 2\pi 1000t + 45°$
15. $v = 5 + 6 \sin 2\pi 100t$

Write the expressions for the following descriptions.

16. A 10 volt peak sine wave at 20 Hz
17. A 5 peak to peak sine wave at 100 Hz with a -1 VDC offset
18. A 10 volt RMS sine wave at 1 kHz lagging by 40 degrees
19. A 20 volt peak sine wave at 10 kHz leading by 20 degrees with a 5 VDC offset

                                         Laboratory Manual for AC Electrical Circuit Analysis

3A

The Oscilloscope (Tektronix MDO 3000 series)

Objective

This exercise is of a particularly practical nature, namely, introducing the use of the oscilloscope. The various input scaling, coupling, and triggering settings are examined along with a few specialty features.

Theory Overview

The oscilloscope (or simply *scope*, for short) is arguably the single most useful piece of test equipment in an electronics laboratory. The primary purpose of the oscilloscope is to plot a voltage versus time although it can also be used to plot one voltage versus another voltage, and in some cases, to plot voltage versus frequency. Oscilloscopes are capable of measuring both AC and DC waveforms, and unlike typical DMMs, can measure AC waveforms of very high frequency (typically 100 MHz or more versus an upper limit of around 1 kHz for a general purpose DMM). It is also worth noting that a DMM will measure the RMS value of an AC sinusoidal voltage, not its peak value.

While the modern digital oscilloscope on the surface appears much like its analog ancestors, the internal circuitry is far more complicated and the instrument affords much greater flexibility in measurement. Modern digital oscilloscopes typically include measurement aides such as horizontal and vertical cursors or bars, as well as direct readouts of characteristics such as waveform amplitude and frequency. At a minimum, modern oscilloscopes offer two input measurement channels although four and eight channel instruments are increasing in popularity.

Unlike handheld DMMs, most oscilloscopes measure voltages with respect to ground, that is, the inputs are not floating and thus the black, or ground, lead is **always** connected to the circuit ground or common node. This is an extremely important point as failure to remember this may lead to the inadvertent short circuiting of components during measurement. The standard accepted method of measuring a non-ground referenced potential is to use two probes, one tied to each node of interest, and then setting the oscilloscope to subtract the two channels rather than display each separately. Note that this technique is not required if the oscilloscope has floating inputs (for example, in a handheld oscilloscope). Further, while it is possible to measure non-ground referenced signals by floating the oscilloscope itself through defeating the ground pin on the power cord, this is a safety violation and should not be done.

Laboratory Manual for AC Electrical Circuit Analysis

Equipment

(1) DC power supply	model:_______________	srn:_________________
(1) AC function generator	model:_______________	srn:_________________
(1) Digital multimeter	model:_______________	srn:_________________
(1) Oscilloscope, Tektronix MDO 3000 series	model:_______________	srn:_________________

Components

(1) 10 kΩ   actual:_________________

(1) 33 kΩ   actual:_________________

Schematics and Diagrams

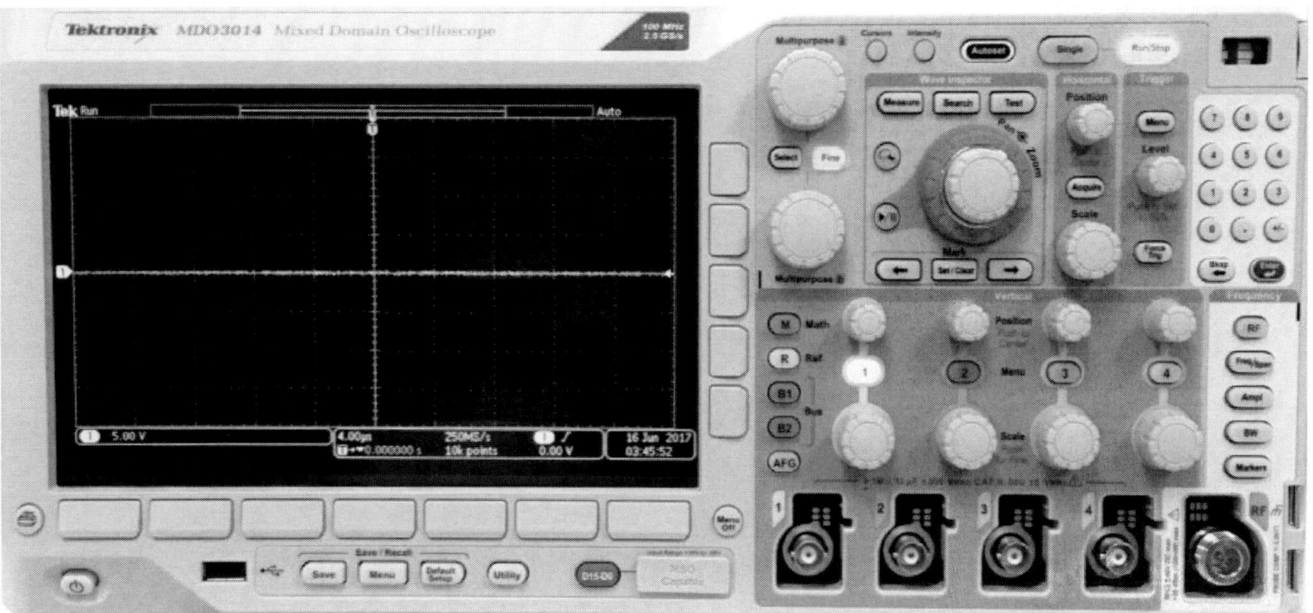

Figure 3A.1

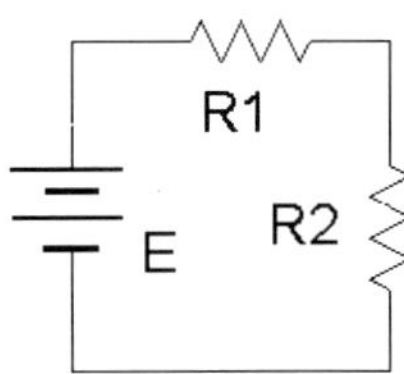

Figure 3A.2

Procedure

1. Figure 3A.1 is a photo of the face of a Tektronix MDO 3000 series oscilloscope. Compare this to the bench oscilloscope and identify the following elements:
   - Channel one through four BNC input connectors.
   - RF input connector and settings section.
   - Channel one through four select buttons.
   - Horizontal Scale (i.e., Sensitivity) and Position knobs.
   - Four Vertical Scale (i.e., Sensitivity) and Position knobs.
   - Trigger Level knob.
   - *Math* and *Measure* (in *Wave Inspector*) buttons.
   - *Save* button (below display).
   - *Autoset* button.
   - *Menu Off* button.

2. Note the numerous buttons along the bottom and side of the display screen. These menu buttons are context-sensitive and their function will depend on the most recently selected button or knob. Menus may be removed from the display by pressing the *Menu Off* button (multiple times for nested menus). Power up the oscilloscope. Note that the main display is similar to a sheet of graph paper. Each square will have an appropriate scaling factor or weighting, for example, 1 volt per division vertically or 2 milliseconds per division horizontally. Waveform voltages and timings may be determined directly from the display by using these scales.

3. Select the channel one and two buttons (yellow and blue) and also press the *Autoset* button. (Autoset tries to create reasonable settings based on the input signal and is useful as a sort of "panic button"). There should now be two horizontal lines on the display, one yellow and one blue. These traces may be moved vertically on the display via the associated Position knobs. Also, a trace can be removed by deselecting the corresponding channel button. The Vertical and Horizontal Scale knobs behave in a similar fashion and **do not** include calibration markings. That is because the settings for these knobs show up on the main display. Adjust the Scale knobs and note how the corresponding values at the bottom of the display change. Voltages are in a 1/2/5 scale sequence while Time is in a 1/2/4 scale sequence.

    Laboratory Manual for AC Electrical Circuit Analysis

4. When an input is selected, a menu will pop up allowing control over that input's basic settings. One of the more important fundamental settings on an oscilloscope channel is the *Input Coupling*. This is controlled via one of the bottom row buttons. There are two choices: *AC* allows only AC signals through thus blocking DC, and *DC* allows **all** signals through (it does **not** prevent AC).

5. Set the channel one Vertical Scale to 5 volts per division. Set the channel two Scale to 2 volts per division. Set the Time (Horizontal) Scale to 1 millisecond per division. Finally, set the input Coupling to DC for both input channels and align the blue and yellow display lines to the center line of the display via the Vertical Position knob (note that pushing the vertical Position knobs will automatically center the trace).

6. Build the circuit of figure 3A.2 using E=10 V, R1=10 kΩ and R2= 33kΩ. Connect a probe from the channel one input to the power supply (red or tip to the positive terminal, black clip to ground). Connect a second probe from channel two to R2 (again, red or tip to the high side of the resistor and the black clip to ground).

7. The yellow and blue lines should have deflected upward. Channel one should be raised two divisions (2 divisions at 5 volts per division yields the 10 volt source). Using this method, determine the voltage across R2 (remember, input two should have been set for 2 volts per division). Calculate the expected voltage across R2 using measured resistor values and compare the two in Table 3A.1. Note that it is not possible to achieve extremely high precision using this method (e.g., four or more digits). Indeed, a DMM is often more useful for direct measurement of DC potentials. Double check the results using a DMM and the final column of Table 3A.1.

8. Select AC Coupling for the two inputs. The flat DC lines should drop back to zero. This is because AC Coupling blocks DC. This will be useful for measuring the AC component of a combined AC/DC signal, such as might be seen in an audio amplifier. Set the input coupling for both channels back to DC.

9. Replace the DC power supply with the function generator. Set the function generator for a one volt peak sine wave at 1 kHz and apply it to the resistor network. The display should now show two small sine waves. Adjust the Vertical Scale settings for the two inputs so that the waves take up the majority of the display. If the display is very blurry with the sine waves appearing to jump about side to side, the Trigger Level may need to be adjusted. Also, adjust the Time Scale so that only one or two cycles of the wave may be seen. Using the Scale settings, determine the two voltages (following the method of step 7) as well as the waveform's period and compare them to the values expected via theory, recording the results in Tables 3A.2 and 3A.3. Also crosscheck the results using a DMM to measure the RMS voltages.

10. To find the voltage across R1, the channel two voltage (V_{R2}) may be subtracted from channel one (E source) via the *Math* function. Use the red button to select the *Math* function and create the appropriate expression from the menu (ch1 – ch2). This display shows up in red. To remove a waveform, press its button again. Remove the math waveform before proceeding to the next step.

11. One of the more useful aspects of the oscilloscope is the ability to show the actual waveshape. This may be used, for example, as a means of determining distortion in an amplifier. Change the waveshape on the function generator to a square wave, triangle, or other shape and note how the oscilloscope responds. Note that the oscilloscope will also show a DC component, if any, as the AC signal being offset or "riding on the DC". Adjust the function generator to add a DC offset to the signal and note how the oscilloscope display shifts. Return the function generator back to a sine wave and remove any DC offset.

12. It is often useful to take precise differential measurement on a waveform. For this, the bars or cursors are useful. Select the *Cursors* button toward the top of the oscilloscope. From the menu on the display, select *Vertical Bars*. Two vertical bars will appear on the display (it is possible that one or both could be positioned off the main display). They may be moved left and right via the Multipurpose knobs (next to the Cursors button). The *Select* button toggles between independent and tandem cursor movement. A read out of the bar values will appear in the upper portion of the display. They indicate the positions of the cursors, i.e., the location where they cross the waveform. Vertical Bars are very useful for obtaining time information as well as amplitudes at specific points along the wave. A similar function is the Horizontal Bars which are particularly useful for determining amplitudes. Try the Horizontal Bars by selecting them via the Cursors menu again (holding the *Cursors* button will bring up the menu).

13. For some waveform parameters, automatic readings are available. These are accessed via the *Measure* button. Press *Measure*, select *Add Measurement*, and page through the various options using the Multipurpose b knob. Select *Frequency*. Note that a small readout of the frequency will now appear on the display. Multiple measurements are possible simultaneously. **Important:** There are specific limits on the proper usage of these measurements. If the guidelines are not followed, erroneous values may result. **Always** perform an approximation via the Scale factor and divisions method even when using an automatic measurement!

14. Finally, a snap-shot of the screen may be saved for future work using the USB port and a USB memory stick via the *Save Menu* button. The pop up menu has options for saving the image as well as the trace data or setup info. Select *Save Screen Image* to save a bit mapped graphics file that can be used as is or processed further in a graphics program (for example, inverting the colors for printing). The .PNG format is recommended.

        Laboratory Manual for AC Electrical Circuit Analysis

Data Tables

V_{R2}	Scale (V/Div)	Number of Divisions	Voltage Scope	Voltage DMM
Oscilloscope				
Theory	X	X		

Table 3.1

	Scale (V/Div)	Number of Divisions	Voltage Peak	Voltage RMS
E Oscilloscope				
E Theory	X	X		
V_{R2} Oscilloscope				
V_{R2} Theory	X	X		

Table 3A.2

	Scale (S/Div)	Number of Divisions	Period	Frequency
E Oscilloscope				
E Theory	X	X		

Table 3A.3

3B

The Oscilloscope (Tektronix TDS 3000 series)

Objective

This exercise is of a particularly practical nature, namely, introducing the use of the oscilloscope. The various input scaling, coupling, and triggering settings are examined along with a few specialty features.

Theory Overview

The oscilloscope (or simply *scope*, for short) is arguably the single most useful piece of test equipment in an electronics laboratory. The primary purpose of the oscilloscope is to plot a voltage versus time although it can also be used to plot one voltage versus another voltage, and in some cases, to plot voltage versus frequency. Oscilloscopes are capable of measuring both AC and DC waveforms, and unlike typical DMMs, can measure AC waveforms of very high frequency (typically 100 MHz or more versus an upper limit of around 1 kHz for a general purpose DMM). It is also worth noting that a DMM will measure the RMS value of an AC sinusoidal voltage, not its peak value.

While the modern digital oscilloscope on the surface appears much like its analog ancestors, the internal circuitry is far more complicated and the instrument affords much greater flexibility in measurement. Modern digital oscilloscopes typically include measurement aides such as horizontal and vertical cursors or bars, as well as direct readouts of characteristics such as waveform amplitude and frequency. At a minimum, modern oscilloscopes offer two input measurement channels although four and eight channel instruments are increasing in popularity.

Unlike handheld DMMs, most oscilloscopes measure voltages with respect to ground, that is, the inputs are not floating and thus the black, or ground, lead is **always** connected to the circuit ground or common node. This is an extremely important point as failure to remember this may lead to the inadvertent short circuiting of components during measurement. The standard accepted method of measuring a non-ground referenced potential is to use two probes, one tied to each node of interest, and then setting the oscilloscope to subtract the two channels rather than display each separately. Note that this technique is not required if the oscilloscope has floating inputs (for example, in a handheld oscilloscope). Further, while it is possible to measure non-ground referenced signals by floating the oscilloscope itself through defeating the ground pin on the power cord, this is a safety violation and should not be done.

                                      Laboratory Manual for AC Electrical Circuit Analysis

Equipment

(1) DC power supply   model:_________________ srn:_________________

(1) AC function generator   model:_________________ srn:_________________

(1) Digital multimeter   model:_________________ srn:_________________

(1) Oscilloscope, Tektronix TDS 3000 series   model:_________________ srn:_________________

Components

(1) 10 kΩ   actual:_________________

(1) 33 kΩ   actual:_________________

Schematics and Diagrams

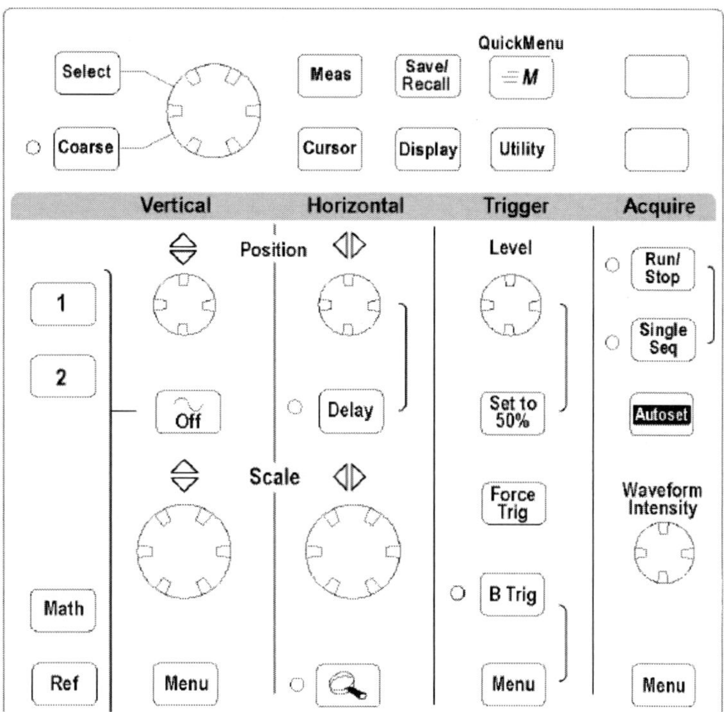

Figure 3B.1

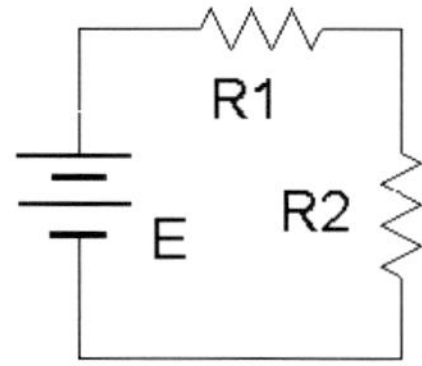

Figure 3B.2

Procedure

1. Figure 3B.1 is an outline of the main face of a Tektronix TDS 3000 series oscilloscope. Compare this to the bench oscilloscope and identify the following elements:
   - Channel one and two BNC input connectors.
   - Trigger BNC input connector.
   - Channel one and two select buttons.
   - Horizontal Sensitivity (or Scale) and Position knobs.
   - Vertical Sensitivity (or Scale) and Position knobs.
   - Trigger Level knob.
   - *Quick Menu* button.
   - *Print/Save* button.
   - *Autoset* button.

2. Note the numerous buttons along the bottom and side of the display screen. These buttons are context-sensitive and their function will depend on the mode of operation of the oscilloscope. Power up the oscilloscope and select the *Quick Menu* button. Notice that the functions are listed next to the buttons. This is a very useful menu and serves as a good starting point for most experiment setups. Note that the main display is similar to a sheet of graph paper. Each square will have an appropriate scaling factor or weighting, for example, 1 volt per division vertically or 2 milliseconds per division horizontally. Waveform voltages and timings may be determined directly from the display by using these scales.

3. Select the channel one and two buttons (yellow and blue) and also select the *Autoset* button. (Autoset tries to create reasonable settings based on the input signal and is useful as a sort of "panic button"). There should now be two horizontal lines on the display, one yellow and one blue. They may be moved via the Position knob. The Position knob moves the currently selected input (select the channel buttons alternately to toggle back and forth between the two inputs). The Vertical and Horizontal Scale knobs behave in a similar fashion and **do not** include calibration markings. That is because the settings for these knobs show up on the main display. Adjust the Scale knobs and note how the corresponding values in the display change. Voltages are in a 1/2/5 scale sequence while Time is in a 1/2/4 scale sequence.

4. One of the more important fundamental settings on an oscilloscope is the *Input Coupling*. This is controlled via one of the bottom row buttons. There are three choices: *Ground* removes the input thus showing a zero reference, *AC* allows only AC signals through thus blocking DC, and *DC* allows **all** signals through (it does **not** prevent AC).

5. Set the channel one Vertical Scale to 5 volts per division. Set the channel two Scale to 2 volts per division. Set the Time (Horizontal) Scale to 1 millisecond per division. Finally, set the input Coupling

to Ground for both input channels and align the blue and yellow display lines to the center line of the display via the Vertical Position knob.

6. Build the circuit of figure 3B.2 using E=10 V, R1=10 kΩ and R2= 33kΩ. Connect a probe from the channel one input to the power supply (red or tip to the positive terminal, black clip to ground). Connect a second probe from channel two to R2 (again, red or tip to the high side of the resistor and the black clip to ground).

7. Switch both inputs to DC coupling. The yellow and blue lines should have deflected upward. Channel one should be raised two divisions (2 divisions at 5 volts per division yields the 10 volt source). Using this method, determine the voltage across R2 (remember, input two should have been set for 2 volts per division). Calculate the expected voltage across R2 using measured resistor values and compare the two in Table 3B.1. Note that it is not possible to achieve extremely high precision using this method (e.g., four or more digits). Indeed, a DMM is often more useful for direct measurement of DC potentials. Double check the results using a DMM and the final column of Table 3B.1.

8. Select AC Coupling for the two inputs. The flat DC lines should drop back to zero. This is because AC Coupling blocks DC. This will be useful for measuring the AC component of a combined AC/DC signal, such as might be seen in an audio amplifier. Set the input coupling for both channels back to DC.

9. Replace the DC power supply with the function generator. Set the function generator for a one volt peak sine wave at 1 kHz and apply it to the resistor network. The display should now show two small sine waves. Adjust the Vertical Scale settings for the two inputs so that the waves take up the majority of the display. If the display is very blurry with the sine waves appearing to jump about side to side, the Trigger Level may need to be adjusted. Also, adjust the Time Scale so that only one or two cycles of the wave may be seen. Using the Scale settings, determine the two voltages (following the method of step 7) as well as the waveform's period and compare them to the values expected via theory, recording the results in Tables 3B.2 and 3B.3. Also crosscheck the results using a DMM to measure the RMS voltages.

10. To find the voltage across R1, the channel two voltage (V_{R2}) may be subtracted from channel one (E source) via the *Math* function. Use the red button to select the *Math* function and create the appropriate expression from the menu (ch1 – ch2). This display shows up in red. To remove a waveform, select it and then select *Off*. Remove the math waveform before proceeding to the next step.

11. One of the more useful aspects of the oscilloscope is the ability to show the actual waveshape. This may be used, for example, as a means of determining distortion in an amplifier. Change the waveshape on the function generator to a square wave, triangle, or other shape and note how the oscilloscope responds. Note that the oscilloscope will also show a DC component, if any, as the AC

signal being offset or "riding on the DC". Adjust the function generator to add a DC offset to the signal and note how the oscilloscope display shifts. Return the function generator back to a sine wave and remove any DC offset.

12. It is often useful to take precise differential measurement on a waveform. For this, the bars or cursors are useful. Select the *Cursor* button toward the top of the oscilloscope. From the menu on the display, select *Vertical Bars*. Two vertical bars will appear on the display (it is possible that one or both could be positioned off the main display). They may be moved left and right via the function knob (next to the Cursor button). The *Select* button toggles between the two cursors. A read out of the bar values will appear in the upper portion of the display. They indicate the positions of the cursors, i.e. the location where they cross the waveform. Vertical Bars are very useful for obtaining time information as well as amplitudes at specific points along the wave. A similar function is the Horizontal Bars which are particularly useful for determining amplitudes. Try the Horizontal Bars by selecting them via the *Cursor* button again.

13. For some waveform parameters, automatic readings are available. These are accessed via the *Meas* (Measurement) button. Select *Meas* and page through the various options. Select *Frequency*. Note that a small readout of the frequency will now appear on the display. Up to four measurements are possible simultaneously. **Important:** There are specific limits on the proper usage of these measurements. If the guidelines are not followed, erroneous values may result. **Always** perform an approximation via the Scale factor and divisions method even when using an automatic measurement!

Data Tables

V_{R2}	Scale (V/Div)	Number of Divisions	Voltage Scope	Voltage DMM
Oscilloscope				
Theory	X	X		

Table 3B.1

	Scale (V/Div)	Number of Divisions	Voltage Peak	Voltage RMS
E Oscilloscope				
E Theory	X	X		
V_{R2} Oscilloscope				
V_{R2} Theory	X	X		

Table 3B.2

	Scale (S/Div)	Number of Divisions	Period	Frequency
E Oscilloscope				
E Theory	X	X		

Table 3B.3

3C

The Oscilloscope (GWInstek 2000)

Objective

This exercise is of a particularly practical nature, namely, introducing the use of the oscilloscope. The various input scaling, coupling, and triggering settings are examined along with a few specialty features.

Theory Overview

The oscilloscope (or simply *scope*, for short) is arguably the single most useful piece of test equipment in an electronics laboratory. The primary purpose of the oscilloscope is to plot a voltage versus time although it can also be used to plot one voltage versus another voltage, and in some cases, to plot voltage versus frequency. Oscilloscopes are capable of measuring both AC and DC waveforms, and unlike typical DMMs, can measure AC waveforms of very high frequency (typically 100 MHz or more versus an upper limit of around 1 kHz for a general purpose DMM). It is also worth noting that a DMM will measure the RMS value of an AC sinusoidal voltage, not its peak value.

While the modern digital oscilloscope on the surface appears much like its analog ancestors, the internal circuitry is far more complicated and the instrument affords much greater flexibility in measurement. Modern digital oscilloscopes typically include measurement aides such as horizontal and vertical cursors or bars, as well as direct readouts of characteristics such as waveform amplitude and frequency. At a minimum, modern oscilloscopes offer two input measurement channels although four and eight channel instruments are increasing in popularity.

Unlike handheld DMMs, most oscilloscopes measure voltages with respect to ground, that is, the inputs are not floating and thus the black, or ground, lead is **always** connected to the circuit ground or common node. This is an extremely important point as failure to remember this may lead to the inadvertent short circuiting of components during measurement. The standard accepted method of measuring a non-ground referenced potential is to use two probes, one tied to each node of interest, and then setting the oscilloscope to subtract the two channels rather than display each separately. Note that this technique is not required if the oscilloscope has floating inputs (for example, in a handheld oscilloscope). Further, while it is possible to measure non-ground referenced signals by floating the oscilloscope itself through defeating the ground pin on the power cord, this is a safety violation and should not be done.

Equipment

(1) DC power supply   model:_________________ srn:_________________

(1) AC function generator   model:_________________ srn:_________________

(1) Digital multimeter   model:_________________ srn:_________________

(1) Oscilloscope, GWInstek 2000 series   model:_________________ srn:_________________

Components

(1) 10 kΩ   actual:_________________

(1) 33 kΩ   actual:_________________

Schematics and Diagrams

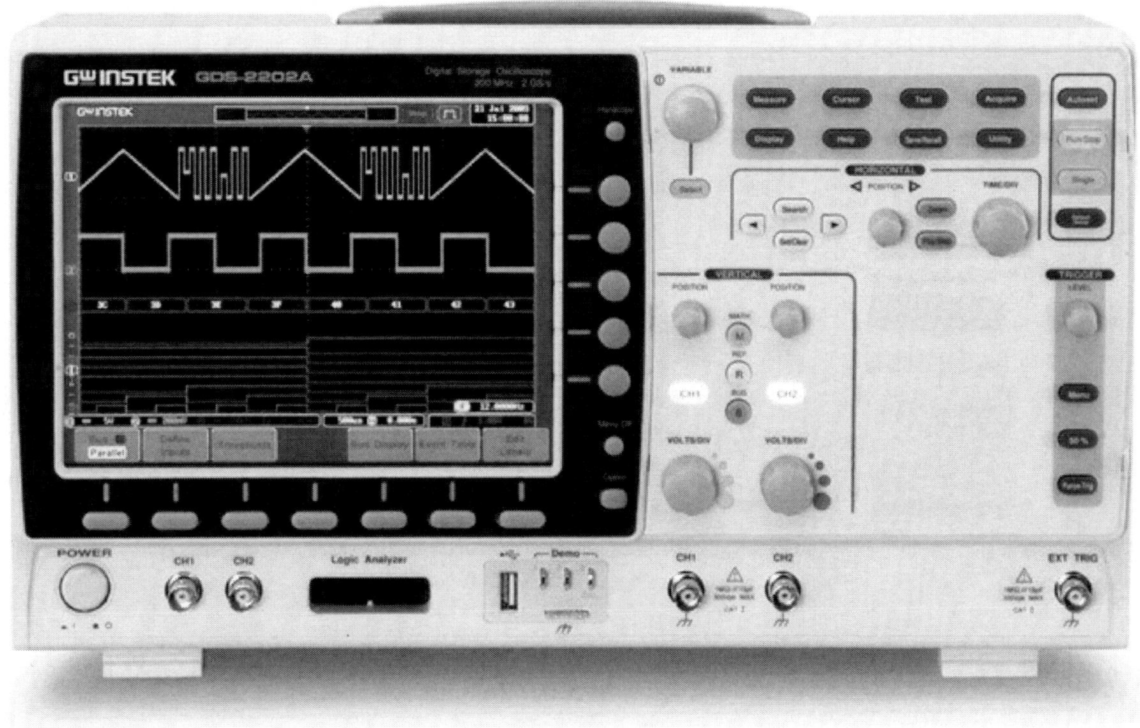

Figure 3C.1A

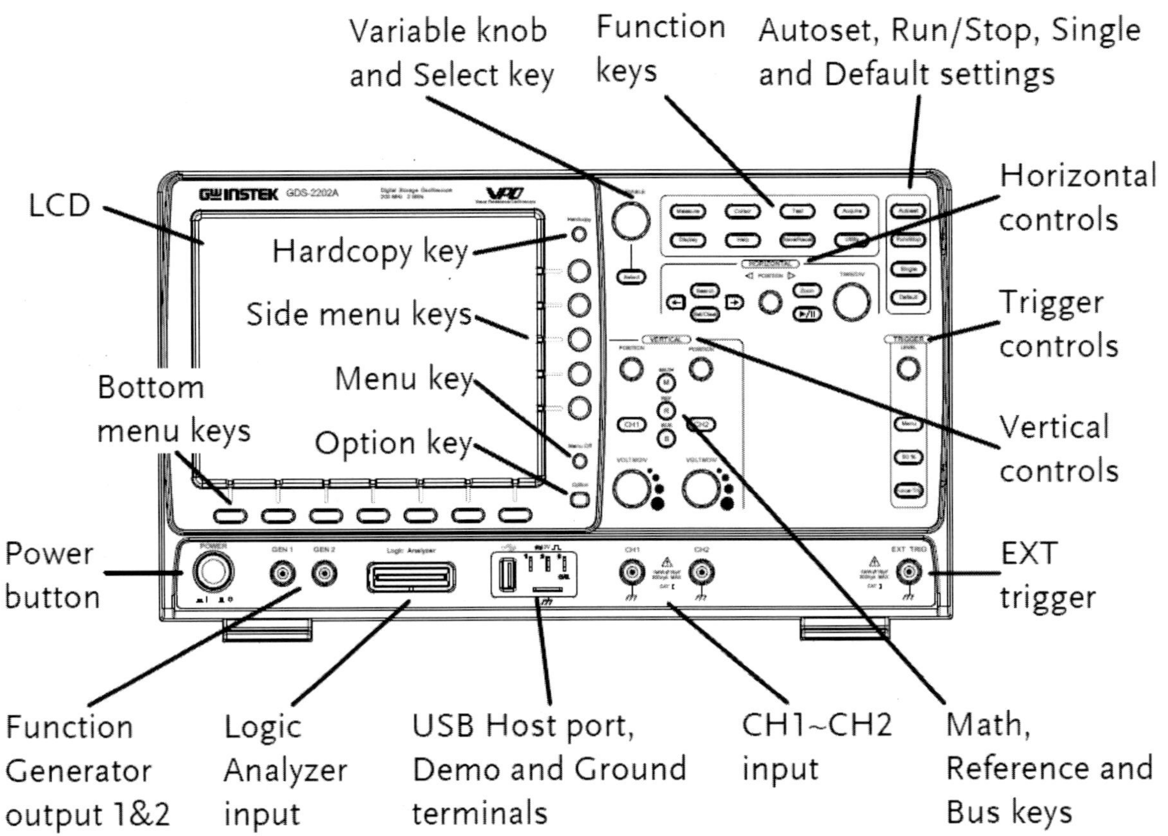

Figure 3C.1B

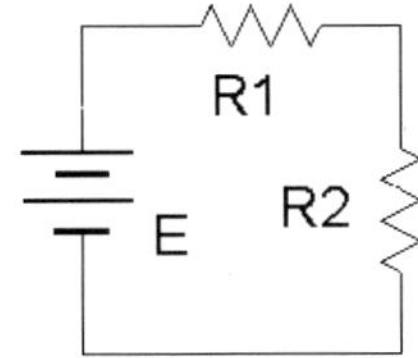

Figure 3C.2

Laboratory Manual for AC Electrical Circuit Analysis

Procedure

1. Figure 3C.1 is an outline of the main face of a GWInstek 2000 series oscilloscope. Compare this to the bench oscilloscope and identify the following elements:
   - Channel one and two BNC input connectors.
   - Trigger BNC input connector.
   - Horizontal sensitivity (Time/Div) and Position knobs.
   - Channel one and two Select buttons (lighted style).
   - Channel one and two Vertical sensitivity (Volts/Div) and Position knobs.
   - Trigger Level knob.
   - *Function* keys, including *Cursor* and *Measure*.
   - *Hardcopy* button.
   - *Autoset* button.
   - *Math* button.
   - USB port.

2. Note the numerous buttons along the bottom and side of the display screen. These buttons are context-sensitive and their function will depend on the mode of operation of the oscilloscope. Power up the oscilloscope. Notice that the functions are listed next to the buttons. Note that the main display is similar to a sheet of graph paper. Each square will have an appropriate scaling factor or weighting, for example, 1 volt per division vertically or 2 milliseconds per division horizontally. Waveform voltages and timings may be determined directly from the display by using these scales.

3. Depress the channel one and two Select buttons (they should light) and also select the *Autoset* button. (Autoset tries to create reasonable settings based on the input signal and is useful as a sort of "panic button"). There should now be two horizontal lines on the display, one yellow and one blue. They may be moved via the Position knob. The Position knobs move the associated input. The Vertical and Horizontal Scale knobs behave in a similar fashion and **do not** include calibration markings. That is because the settings for these knobs show up on the main display. Adjust the Scale knobs and note how the corresponding values in the display change. Voltages and Time base use a 1/2/5 scale sequence.

4. One of the more important fundamental settings on an oscilloscope is the *Input Coupling*. This is controlled via one of the bottom row buttons. There are three choices: *Ground* removes the input thus showing a zero reference, *AC* allows only AC signals through thus blocking DC, and *DC* allows **all** signals through (it does **not** prevent AC).

5. Set the channel one Vertical Scale to 5 volts per division. Set the channel two Scale to 2 volts per division. Set the Time (Horizontal) Scale to 1 millisecond per division. Finally, set the input Coupling to Ground for both input channels and align the blue and yellow display lines to the center line of the display via the Vertical Position knobs.

6. Build the circuit of Figure 3C.2 using E=10 V, R1=10 kΩ and R2= 33kΩ. Connect a probe from the channel one input to the power supply (red or tip to plus, black clip to ground). Connect a second probe from channel two to R2 (again, red or tip to the high side of the resistor and the black clip to ground).

7. Switch both inputs to DC coupling. The yellow and blue lines should have deflected upward. Channel one should be raised two divisions (2 divisions times 5 volts per division yields the 10 volt source). Using this method, determine the voltage across R2 (remember, input two should have been set for 2 volts per division). Calculate the expected voltage across R2 using measured resistor values and compare the two in Table 3C.1. Note that it is not possible to achieve extremely high precision using this method (e.g., four or more digits). Indeed, a DMM is often more useful for direct measurement of DC potentials. Double check the results using a DMM and the final column of Table 3C.1.

8. Select AC Coupling for the two inputs. The flat DC lines should drop back to zero. This is because AC Coupling blocks DC. This will be useful for measuring the AC component of a combined AC/DC signal, such as might be seen in an audio amplifier. Set the input coupling for both channels back to DC.

9. Replace the DC power supply with the function generator. Set the function generator for a one volt peak sine wave at 1 kHz and apply it to the resistor network. The display should now show two small sine waves. Adjust the Vertical Scale settings for the two inputs so that the waves take up the majority of the display. If the display is very blurry with the sine waves appearing to jump about side to side, the Trigger Level may need to be adjusted. Also, adjust the Time Scale so that only one or two cycles of the wave may be seen. Using the Scale settings, determine the two voltages (following the method of step 7) as well as the waveform's period and compare them to the values expected via theory, recording the results in Tables 3C.2 and 3C.3. Also crosscheck the results using a DMM to measure the RMS voltages.

10. To find the voltage across R1, the channel two voltage (V_{R2}) may be subtracted from channel one (E source) via the *Math* function. Use the red button to select the *Math* function and create the appropriate expression from the menu (ch1 – ch2). This display shows up in red. To remove a waveform, simply deselect it (depress the associated button). Remove the math waveform before proceeding to the next step.

11. One of the more useful aspects of the oscilloscope is the ability to show the actual waveshape. This may be used, for example, as a means of determining distortion in an amplifier. Change the waveshape on the function generator to a square wave, triangle, or other shape and note how the oscilloscope responds. Note that the oscilloscope will also show a DC component, if any, as the AC signal being offset or "riding on the DC". Adjust the function generator to add a DC offset to the

signal and note how the oscilloscope display shifts. Return the function generator back to a sine wave and remove any DC offset.

12. It is often useful to take precise differential measurement on a waveform. For this, the bars or cursors are useful. Select the *Cursor* button toward the top of the oscilloscope. From the menu on the display, select *Vertical*. Two vertical bars will appear on the display (it is possible that one or both could be positioned off the main display). They may be moved left and right via the *Variable* knob (next to the Cursor button). The *Select* button toggles between the two cursors. A read out of the bar values will appear in the upper portion of the display. They indicate the positions of the cursors, i.e. the location where they cross the waveform. Vertical Bars are very useful for obtaining time information as well as amplitudes at specific points along the wave. A similar function is the Horizontal Bars which are particularly useful for determining amplitudes. Try the Horizontal Bars by selecting them via the *Cursor* button again.

13. For some waveforms parameters, automatic readings are available. These are accessed via the *Measure* button. Select *Measure* and page through the various options. Select *Frequency*. Note that a small readout of the frequency will now appear on the display. Now try *RMS* and compare the result to that given by the DMM earlier. Note that several measurements are possible simultaneously. **Important:** There are specific limits on the proper usage of these measurements. If the guidelines are not followed, erroneous values may result. **Always** perform an approximation via the Scale factor and divisions method even when using an automatic measurement!

14. Finally, a snap-shot of the screen may be saved for future work using the USB port and a USB memory stick via the *Hardcopy* button. The result will be a bit mapped graphics file that can be used as is (see below) or processed further in a graphics program (for example, inverting the colors for printing).

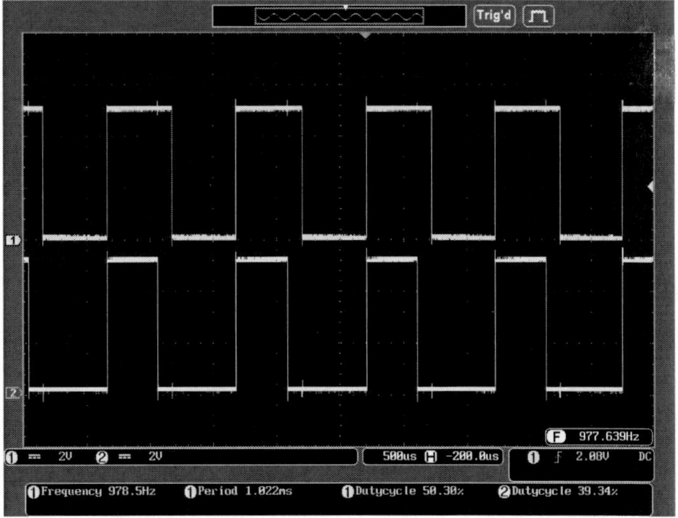

Figure 3

Data Tables

V_{R2}	Scale (V/Div)	Number of Divisions	Voltage Scope	Voltage DMM
Oscilloscope				
Theory	X	X		

Table 1

	Scale (V/Div)	Number of Divisions	Voltage Peak	Voltage RMS
E Oscilloscope				
E Theory	X	X		
V_{R2} Oscilloscope				
V_{R2} Theory	X	X		

Table 2

	Scale (S/Div)	Number of Divisions	Period	Frequency
E Oscilloscope				
E Theory	X	X		

Table 3

4

Capacitive Reactance

Objective

Capacitive reactance will be examined in this exercise. In particular, its relationship to capacitance and frequency will be investigated, including a plot of capacitive reactance versus frequency.

Theory Overview

The current – voltage characteristic of a capacitor is unlike that of typical resistors. While resistors show a constant resistance value over a wide range of frequencies, the equivalent ohmic value for a capacitor, known as *capacitive reactance*, is inversely proportional to frequency. The capacitive reactance may be computed via the formula:

$$X_c = -j\frac{1}{2\pi f C}$$

The magnitude of capacitive reactance may be determined experimentally by feeding a capacitor a known current, measuring the resulting voltage, and dividing the two, following Ohm's law. This process may be repeated a across a range of frequencies in order to obtain a plot of capacitive reactance versus frequency. An AC current source may be approximated by placing a large resistance in series with an AC voltage, the resistance being considerably larger than the maximum reactance expected.

Equipment

(1) AC function generator	model:_______________	srn:_________________
(1) Oscilloscope	model:_______________	srn:_________________

Components

(1) 1 μF	actual:_________________
(1) 2.2 μF	actual:_________________
(1) 10 kΩ	actual:_________________

Schematics

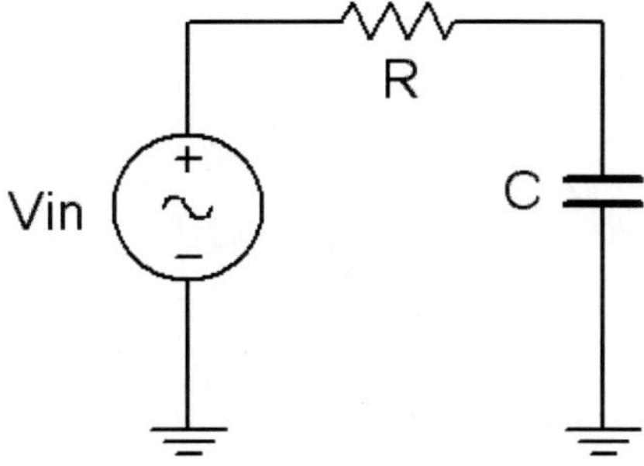

Figure 4.1

Procedure

Current Source

1. Using figure 4.1 with Vin=10 V p-p and R=10 kΩ, and assuming that the reactance of the capacitor is much smaller than 10k and can be ignored, determine the circulating current using measured component values and record in Table 4.1.

Measuring Reactance

2. Build the circuit of figure 4.1 using R=10 kΩ, and C=1 μF. Place one probe across the generator and another across the capacitor. Set the generator to a 200 Hz sine wave and 10 V p-p. Make sure that the *Bandwidth Limit* of the oscilloscope is engaged for both channels. This will reduce the signal noise and make for more accurate readings.

3. Calculate the theoretical value of Xc using the measured capacitor value and record in Table 4.2.

4. Record the peak-to-peak capacitor voltage and record in Table 4.2.

5. Using the source current from Table 4.1 and the measured capacitor voltage, determine the experimental reactance and record it in Table 4.2. Also compute and record the deviation.

6. Repeat steps three through five for the remaining frequencies of Table 4.2.

7. Replace the 1 μF capacitor with the 2.2 μF unit and repeat steps two through six, recording results in Table 4.3.

8. Using the data of Tables 4.2 and 4.3, create plots of capacitive reactance versus frequency.

Data Tables

$i_{source\ (p\text{-}p)}$	

Table 4.1

Frequency	X_C Theory	$V_{C(p\text{-}p)}$ Exp	X_C Exp	% Dev
200				
400				
600				
800				
1.0 k				
1.2 k				
1.6 k				
2.0 k				

Table 4.2

Frequency	X_C Theory	$V_{C(p\text{-}p)}$ Exp	X_C Exp	% Dev
200				
400				
600				
800				
1.0 k				
1.2 k				
1.6 k				
2.0 k				

Table 4.3

Questions

1. What is the relationship between capacitive reactance and frequency?

2. What is the relationship between capacitive reactance and capacitance?

3. If the experiment had been repeated with frequencies 10 times higher than those in Table 4.2, what would the resulting plots look like?

4. If the experiment had been repeated with frequencies 10 times lower than those in Table 4.2, what effect would that have on the experiment?

5

Inductive Reactance

Objective

Inductive reactance will be examined in this exercise. In particular, its relationship to inductance and frequency will be investigated, including a plot of inductive reactance versus frequency.

Theory Overview

The current – voltage characteristic of an inductor is unlike that of typical resistors. While resistors show a constant resistance value over a wide range of frequencies, the equivalent ohmic value for an inductor, known as *inductive reactance*, is directly proportional to frequency. The inductive reactance may be computed via the formula:

$$X_L = j2\pi f L$$

The magnitude of inductive reactance may be determined experimentally by feeding an inductor a known current, measuring the resulting voltage, and dividing the two, following Ohm's law. This process may be repeated a across a range of frequencies in order to obtain a plot of inductive reactance versus frequency. An AC current source may be approximated by placing a large resistance in series with an AC voltage, the resistance being considerably larger than the maximum reactance expected.

Equipment

(1) AC function generator	model:_______________	srn:_______________
(1) Oscilloscope	model:_______________	srn:_______________
(1) DMM	model:_______________	srn:_______________

Components

(1) 1 mH	actual:_______________
(1) 10 mH	actual:_______________
(1) 10 kΩ	actual:_______________

Schematics

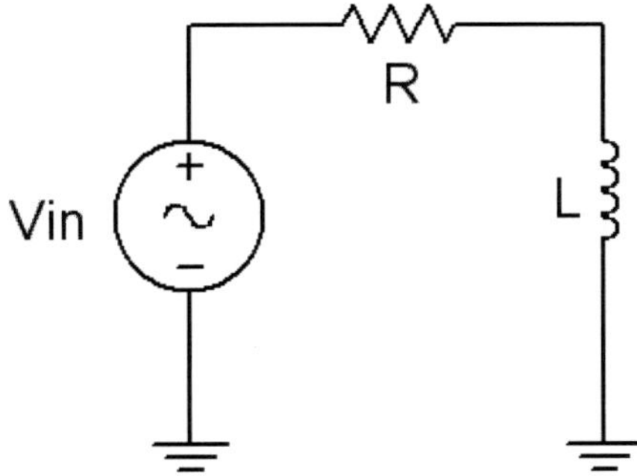

Figure 5.1

Procedure

Current Source

1. Using figure 5.1 with Vin=10 V p-p and R=10 kΩ, and assuming that the reactance of the inductor is much smaller than 10k and can be ignored, determine the circulating current using measured component values and record in Table 5.1. Also, measure the DC coil resistances of the inductors using an ohmmeter or DMM and record in Table 5.1.

Measuring Reactance

2. Build the circuit of figure 5.1 using R=10 kΩ, and L=10 mH. Place one probe across the generator and another across the inductor. Set the generator to a 1000 Hz sine wave and 10 V p-p. Make sure that the *Bandwidth Limit* of the oscilloscope is engaged for both channels. This will reduce the signal noise and make for more accurate readings.

3. Calculate the theoretical value of X_L using the measured inductor value and record in Table 5.2.

4. Record the peak-to-peak inductor voltage and record in Table 5.2.

5. Using the source current from Table 5.1 and the measured inductor voltage, determine the experimental reactance and record it in Table 5.2. Also compute and record the deviation.

6. Repeat steps three through five for the remaining frequencies of Table 5.2.

7. Replace the 10 mH inductor with the 1 mH unit and repeat steps two through six, recording results in Table 5.3.

8. Using the data of Tables 5.2 and 5.3, create plots of inductive reactance versus frequency.

Data Tables

$i_{source(p\text{-}p)}$	
R_{coil} of 10 mH	
R_{coil} of 1 mH	

Table 5.1

Frequency	X_L Theory	$V_{L(p\text{-}p)}$ Exp	X_L Exp	% Dev
1 k				
2 k				
3 k				
4 k				
5 k				
6 k				
8 k				
10 k				

Table 5.2

                    Laboratory Manual for AC Electrical Circuit Analysis

Frequency	X_L Theory	$V_{L(p\text{-}p)}$ Exp	X_L Exp	% Dev
10 k				
20 k				
30 k				
40 k				
50 k				
60 k				
80 k				
100 k				

Table 5.3

Questions

1. What is the relationship between inductive reactance and frequency?

2. What is the relationship between inductive reactance and inductance?

3. If the 10 mH trial had been repeated with frequencies 10 times higher than those in Table 5.2, what effect would that have on the experiment?

4. Do the coil resistances have any effect on the plots?

Series R, L, C Circuits

Objective

This exercise examines the voltage and current relationships in series R, L, C networks. Of particular importance is the phase of the various components and how Kirchhoff's voltage law is extended for AC circuits. Both time domain and phasor plots of the voltages are generated.

Theory Overview

Each element has a unique phase response: for resistors, the voltage is always in phase with the current, for capacitors the voltage always lags the current by 90 degrees, and for inductors the voltage always leads the current by 90 degrees. Consequently, a series combination of R, L, and C components will yield a complex impedance with a phase angle between +90 and -90 degrees. Due to the phase response, Kirchhoff's voltage law must be computed using vector (phasor) sums rather than simply relying on the magnitudes. Indeed, all computations of this nature, such as a voltage divider, must be computed using vectors.

Equipment

(1) AC function generator	model:_______________	srn:_______________
(1) Oscilloscope	model:_______________	srn:_______________

Components

(1) 10 nF	actual:_______________
(1) 10 mH	actual:_______________
(1) 1 kΩ	actual:_______________

Schematics

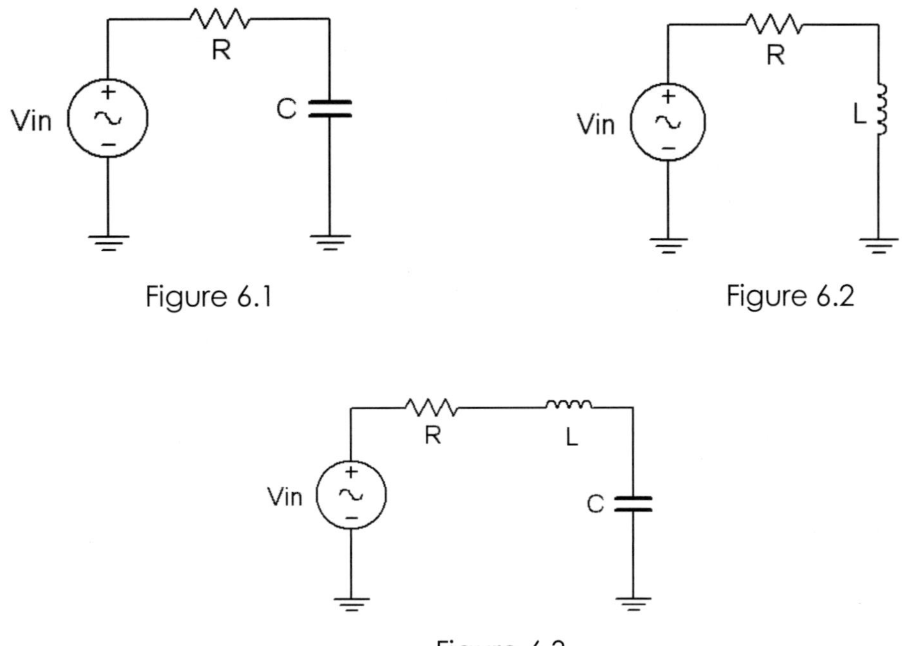

Figure 6.1   Figure 6.2

Figure 6.3

Procedure

RC Circuit

1. Using Figure 6.1 with Vin=2 V p-p sine at 10 kHz, R=1 kΩ, and C=10 nF, determine the theoretical capacitive reactance and circuit impedance, and record the results in Table 6.1 (the experimental portion of this table will be filled out in step 5). Using the voltage divider rule, compute the resistor and capacitor voltages and record them in Table 6.2.

2. Build the circuit of Figure 6.1 using R=1 kΩ, and C=10 nF. Place one probe across the generator and another across the capacitor. Set the generator to a 10 kHz sine wave and 2 V p-p. Make sure that the *Bandwidth Limit* of the oscilloscope is engaged for both channels. This will reduce the signal noise and make for more accurate readings. Also, consider using *Averaging* for the acquisition mode, particularly to clean up signals derived using the *Math* function.

3. Measure the peak-to-peak voltage across the capacitor and record in Table 6.2. Along with the magnitude, be sure to record the time deviation between V_C and the input signal (from which the phase may be determined). Using the *Math* function, measure and record the voltage and time delay for the resistor ($V_{in} - V_C$). Compute the phase angle and record these values in Table 6.2.

4. Take a snapshot of the oscilloscope displaying V_{in}, V_C, and V_R.

5. Compute the deviations between the theoretical and experimental values of Table 6.2 and record the results in the final columns of Table 6.2. Based on the experimental values, determine the experimental Z and X_C values via Ohm's law ($i=V_R/R$, $X_C=V_C/i$, $Z=V_{in}/i$) and record back in Table 6.1 along with the deviations.

6. Create a phasor plot showing V_{in}, V_C, and V_R. Include both the time domain display from step 4 and the phasor plot with the technical report.

RL Circuit

7. Replace the capacitor with the 10 mH inductor (i.e. Figure 6A.2), and repeat steps 1 through 6 in like manner, using Tables 6.3 and 6.4.

RLC Circuit

8. Using Figure 6.3 with both the 10 nF capacitor and 10 mH inductor, repeat steps 1 through 6 in similar manner, using Tables 6.5 and 6.6. **Using a four channel oscilloscope:** To obtain proper readings, place the first probe at the input, the second probe between the resistor and inductor, and the third probe between the inductor and capacitor. Probe three yields V_C. Using the *Math* function, probe two minus probe three yields V_L, and finally, probe one minus probe two yields V_R. Assigning *Reference* waveforms can be useful to see all of the signals together. **Using a two channel oscilloscope:** Unfortunately, it will be impossible to see the voltage of all three components simultaneously with the source voltage using a two channel oscilloscope. To obtain proper readings, place the first probe at the input and the second probe across the capacitor in order to see the phase and magnitude of V_C. Then, swap C and L (placing the second probe across the inductor) to see V_L, and finally, swap L and R (with the second probe across R) in order see V_R.

Data Tables

RC Circuit

	Theory	Experimental	% Deviation
X_C			
Z Magnitude			
Z θ			

Table 6.1

	Theory Mag	Theory θ	Exp Mag	Exp Delay	Exp θ	% Dev Mag	% Dev θ
V_C							
V_R							

Table 6.2

RL Circuit

	Theory	Experimental	% Deviation
X_L			
Z Magnitude			
Z θ			

Table 6.3

	Theory Mag	Theory θ	Exp Mag	Exp Delay	Exp θ	% Dev Mag	% Dev θ
V_L							
V_R							

Table 6.4

RLC Circuit

	Theory	Experimental	% Deviation
X_C			
X_L			
Z Magnitude			
Z θ			

Table 6.5

	Theory Mag	Theory θ	Exp Mag	Exp Delay	Exp θ	% Dev Mag	% Dev θ
V_C							
V_L							
V_R							

Table 6.6

Questions

1. What is the phase relationship between R, L, and C components in a series AC circuit?

2. Based on measurements, does Kirchhoff's voltage law apply to the three tested circuits (show work)?

3. In general, how would the phasor diagram of Figure 6.1 change if the frequency was raised?

4. In general, how would the phasor diagram of Figure 6.2 change if the frequency was lowered?

Laboratory Manual for AC Electrical Circuit Analysis

7

Parallel R, L, C Circuits

Objective

This exercise examines the voltage and current relationships in parallel R, L, C networks. Of particular importance is the phase of the various components and how Kirchhoff's current law is extended for AC circuits. Both time domain and phasor plots of the currents are generated. A technique to measure current using a *current sense resistor* will also be explored.

Theory Overview

Recall that for resistors, the voltage is always in phase with the current, for capacitors the voltage always lags the current by 90 degrees, and for inductors the voltage always leads the current by 90 degrees. Because each element has a unique phase response between +90 and -90 degrees, a parallel combination of R, L, and C components will yield a complex impedance with a phase angle between +90 and -90 degrees. Due to the phase response, Kirchhoff's current law must be computed using vector (phasor) sums rather than simply relying on the magnitudes. Indeed, all computations of this nature, such as a current divider, must be computed using vectors.

Equipment

(1) AC function generator	model:_______________	srn:_______________
(1) Oscilloscope	model:_______________	srn:_______________

Components

(1) 10 nF	actual:_______________
(1) 10 mH	actual:_______________
(1) 1 kΩ	actual:_______________
(3) 10 Ω	actual:_______________
	actual:_______________
	actual:_______________

Schematics

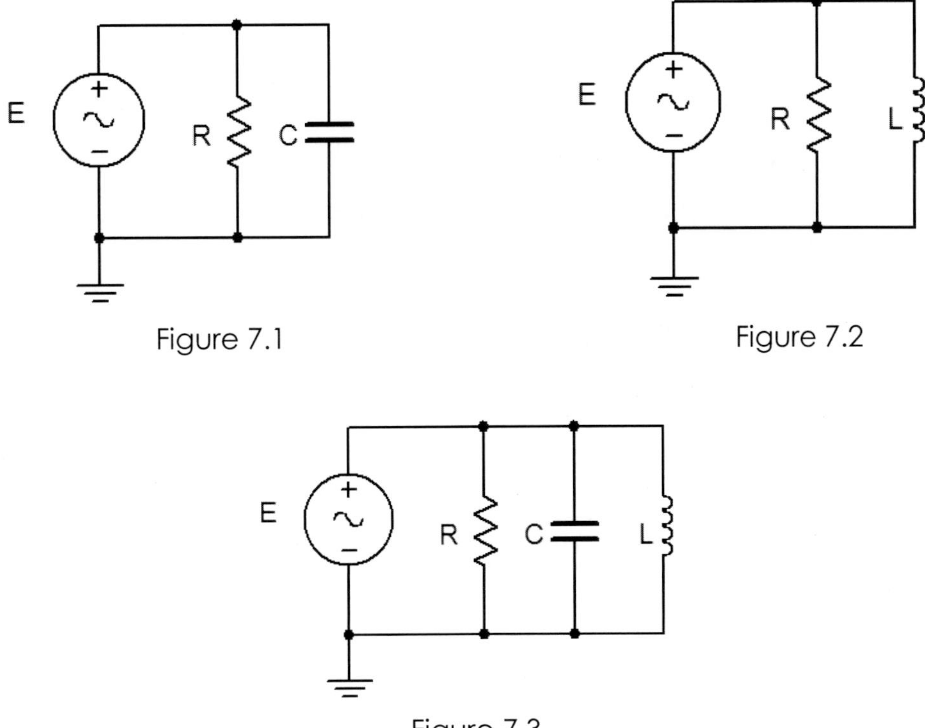

Figure 7.1

Figure 7.2

Figure 7.3

Procedure

RC Circuit

1. Using Figure 7.1 with a 10 V p-p 10 kHz source, R=1 kΩ, and C=10 nF, determine the theoretical capacitive reactance and circuit impedance, and record the results in Table 7.1 (the experimental portion of this table will be filled out in step 6). Using the current divider rule, compute the resistor and capacitor currents and record them in Table 7.2.

2. Build the circuit of Figure 7.1 using R=1 kΩ, and C=10 nF. A common method to measure current using the oscilloscope is to place a small *current sense resistor* in line with the current of interest. If the resistor is much smaller than the surrounding reactances it will have a minimal effect on the current. Because the voltage and current of the resistor are always in phase with each other, the relative phase of the current in question must be the same as that of the sensing resistor's voltage. Each of the three circuit currents will be measured separately and with respect to the source in order to determine relative phase. To measure the total current, place a 10 Ω resistor between ground and the bottom connection of the parallel components. Set the generator to a 10 V p-p sine wave at 10 kHz. Make sure that the *Bandwidth Limit* of the oscilloscope is engaged for both channels. This will reduce the signal noise and make for more accurate readings. Also, consider using waveform averaging, particularly to clean up signals derived via the *Math* function.

3. Place probe one across the generator and probe two across the sense resistor. Measure the voltage across the sense resistor, calculate the corresponding total current via Ohm's law and record in Table 7.2. Along with the magnitude, be sure to record the time deviation between the sense waveform and the input signal (from which the phase may be determined eventually).

4. Remove the main sense resistor and place one 10 Ω resistor between the capacitor and ground to serve as the capacitor current sense. Place a second 10 Ω resistor between the resistor and ground to sense the resistor current. Leave probe one at the generator and move probe two across the sense resistor in the resistor branch. Repeat the Ohm's law process to obtain its current, recording the magnitude and phase angle in Table 7.2. Finally, move probe two so that it is across the capacitor's sense resistor. Measure and record the appropriate values in Table 7.2. Note that if you are using a four channel oscilloscope, simultaneous input, resistor and capacitor measurements are possible.

5. Move probe one to the resistor's sense resistor and leave probe two at the capacitor's sense resistor. Save a picture of the oscilloscope displaying the voltage waveforms representing i_R, i_C and i_{in} (i.e., the *Math* waveform computed from $i_R + i_C$).

6. Compute the deviations between the theoretical and experimental values of Table 7.2 and record the results in the final columns of Table 7.2. Based on the experimental values, determine the experimental Z and X_C values via Ohm's law ($X_C = V_C/i_C$, $Z = V_{in}/i_{in}$) and record back in Table 7.1 along with the deviations.

7. Create a phasor plot showing i_{in}, i_C, and i_R. Include both the time domain display from step 4 and the phasor plot with the technical report.

RL Circuit

8. Replace the capacitor with the 10 mH inductor (i.e. Figure 7A.2), and repeat steps 1 through 7 in like manner, using Tables 7.3 and 7.4.

RLC Circuit

9. Using Figure 7.3 with both the 10 nF capacitor and 10 mH inductor (and a third sense resistor), repeat steps 1 through 7 in like manner, using Tables 7.5 and 7.6. Note that it will not be possible to see all four waveforms simultaneously in step 5 if a **two channel oscilloscope** is being used. For a **four channel oscilloscope**, place a probe across each of the three sense resistors.

Data Tables

RC Circuit

	Theory	Experimental	% Deviation
X_C			
Z Magnitude			
Z θ			

Table 7.1

	Theory Mag	Theory θ	Exp Mag	Exp Delay	Exp θ	% Dev Mag	% Dev θ
i_C							
i_R							
i_{in}							

Table 7.2

RL Circuit

	Theory	Experimental	% Deviation
X_L			
Z Magnitude			
Z θ			

Table 7.3

	Theory Mag	Theory θ	Exp Mag	Exp Delay	Exp θ	% Dev Mag	% Dev θ
i_L							
i_R							
i_{in}							

Table 7.4

RLC Circuit

	Theory	Experimental	% Deviation
X_C			
X_L			
Z Magnitude			
Z θ			

Table 7.5

	Theory Mag	Theory θ	Exp Mag	Exp Delay	Exp θ	% Dev Mag	% Dev θ
i_C							
i_L							
i_R							
i_{in}							

Table 7.6

Questions

1. What is the phase relationship between R, L, and C components in a parallel AC circuit?

2. Based on measurements, does Kirchhoff's current law apply to the three tested circuits (show work)?

3. In general, how would the phasor diagram of Figure 7.1 change if the frequency was raised?

4. In general, how would the phasor diagram of Figure 7.2 change if the frequency was lowered?

     Laboratory Manual for AC Electrical Circuit Analysis

8

Series-Parallel R, L, C Circuits

Objective

This exercise examines the voltage and current relationships in series-parallel R, L, C networks. Often series-parallel circuits may be analyzed along the lines of the simpler series-only or parallel-only circuits, but where each "element" may comprise a complex impedance rather than a singular R, L, or C component. Both Kirchhoff's current law and Kirchhoff's voltage law may be applied to these circuits. In this exercise, both time domain and phasor plots of the voltages and currents are generated.

Theory Overview

Many complex R, L, C networks may be analyzed by reducing them to simpler series or parallel circuits, perhaps through an iterative process in more involved instances. In this analysis each series or parallel element is in fact a complex impedance made up of a series or parallel combination of other components and thus producing a phase angle between +90 and -90 degrees. Consequently, the simple "all right angles" phasor diagrams found for basic series and parallel circuits may be replaced with more general phasor diagrams with non-right angles. In spite of this, both Kirchhoff's current law and Kirchhoff's voltage law must still be satisfied for the entire circuit, as well as any sub-circuits or branches.

Equipment

(1) AC function generator	model:____________	srn:____________
(1) Oscilloscope	model:____________	srn:____________

Components

(1) 10 nF	actual:____________
(1) 10 mH	actual:____________
(1) 1 kΩ	actual:____________
(2) 10 Ω	actual:____________

Schematics

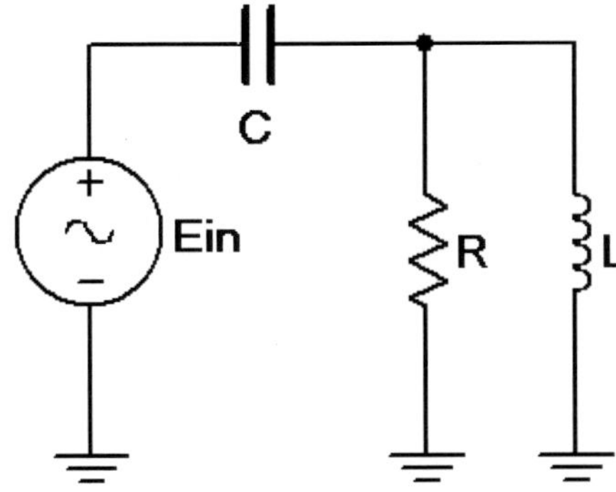

Figure 8.1

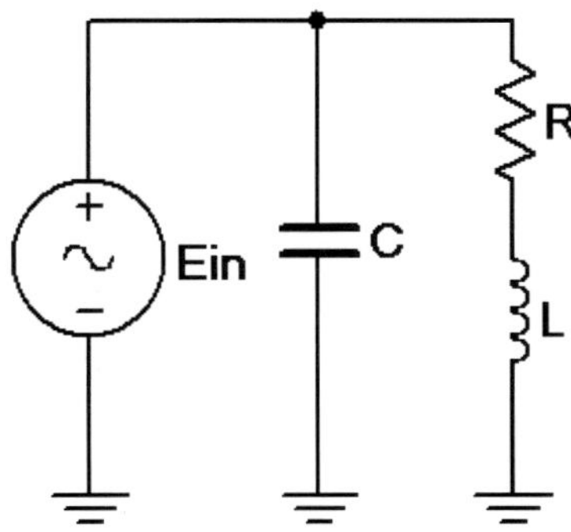

Figure 8.2

Procedure

Circuit 1

1. Using Figure 8.1 with a 10 kHz sine wave at 10 V p-p source, R=1 kΩ, L=10 mH, and C=10 nF, determine the theoretical inductive and capacitive reactances, parallel branch reactance and total circuit impedance, and record the results in Table 8.1 (the experimental portion of this table will be filled out in step 5). Using Ohm's law and the voltage divider rule compute the capacitor and inductor-resistor voltages along with the input current and record them in Table 8.2.

2. Build the circuit of Figure 8.1 using R=1 kΩ, L=10 mH, and C=10 nF. Set the generator to a 10 kHz sine wave and 10 V p-p. Make sure that the *Bandwidth Limit* of the oscilloscope is engaged for both channels. This will reduce the signal noise and make for more accurate readings.

3. Place probe one across the generator and probe two across the parallel inductor-resistor branch. Using the *Math* function, the capacitor voltage may be found by subtracting the voltage of probe two from that of probe one. Also, the input current may be found by dividing the capacitor's voltage by its reactance. Measure the parallel branch voltage and capacitor voltage, both magnitude and phase, and record in Table 8.2. Compute the input current and record in Table 8.2.

4. Take a picture of the three voltage waveforms.

5. Compute the deviations between the theoretical and experimental values of Table 8.2 and record the results in the final columns of Table 8.2. Based on the experimental values, determine the experimental total Z and parallel branch Z values via Ohm's law (e.g., $Z_T=V_{in}/i_{in}$) and record back in Table 8.1 along with the deviations.

6. Create a phasor plot showing V_{in}, V_{LR}, and V_C. Include both the time domain display from step 4 and the phasor plot with the technical report.

Circuit 2

7. Using Figure 8.2 with a 10k Hz sine wave at 10 V p-p, R=1 kΩ, L=10 mH, and C=10 nF, determine the theoretical inductive and capacitive reactances, series branch impedance and total circuit impedance, and record the results in Table 8.3. Using Ohm's law compute the capacitor and inductor-resistor currents along with the input current and record them in Table 8.4.

8. Build the circuit of Figure 8.2 using R=1 kΩ, L=10 mH, and C=10 nF. Insert a 10 Ω current sense resistor at the bottom of the LR leg and another at the bottom of the capacitor leg. Set the generator to a 10 kHz sine wave and 10 V p-p. Make sure that the *Bandwidth Limit* of the oscilloscope is engaged for both channels. This will reduce the signal noise and make for more accurate readings.

9. Place probe one across the generator and probe two across the inductor-resistor branch sense resistor. The inductor-resistor current may be found by dividing the probe two voltage by the sense resistor. The capacitor current is found in a similar manner using its current sense resistor (use probe three if available, otherwise perform this twice using probe two). Record both magnitude and phase of the two currents in Table 8.4.

10. Take a picture of the V_{in} and i_{LR} sense waveforms and also of the V_{in} and i_C sense waveforms (one combined picture if using three probes, otherwise two separate pictures).

11. To measure the input current, remove the two sense resistors and place one of them so that it is between ground and the bottom junction of the resistor and capacitor. Move probe two to this sense resistor and measure the voltage. From this, compute the total current and record both magnitude and phase in Table 8.4.

12. Take a picture of the V_{in} and i_{in} sense waveforms.

13. Compute the deviations between the theoretical and experimental values of Table 8.4 and record the results in the final columns of Table 8.4. Based on the experimental values, determine the experimental total Z and series branch Z values and record back in Table 8.3 along with the deviations.

14. Create a phasor plot showing i_{in}, i_{LR}, and i_C. Include both the time domain displays from steps 10 & 12 and the phasor plot with the technical report.

Computer Simulation

15. Build the circuit of Figure 8.1 in a simulator. Using Transient Analysis, determine the voltage across the inductor and compare the magnitude and phase to the theoretical and measured values recorded in Table 8.2.

Data Tables

Circuit 1

	Theory	Experimental	% Deviation
X_C			
X_L			
$R \mid \mid X_L$			
Z_T Magnitude			
$Z_T \theta$			

Table 8.1

	Theory Mag	Theory θ	Exp Mag	Exp Delay	Exp θ	% Dev Mag	% Dev θ
V_{LR}							
V_C							
i_{in}							

Table 8.2

Circuit 2

	Theory	Experimental	% Deviation
X_C			
X_L			
$R + X_L$			
Z_T Magnitude			
Z_T θ			

Table 8.3

	Theory Mag	Theory θ	Exp Mag	Exp Delay	Exp θ	% Dev Mag	% Dev θ
i_{LR}							
i_C							
i_{in}							

Table 8.4

Questions

1. Is the phase relationship between circuit voltages or currents in a series-parallel AC circuit necessarily a right-angle relationship?

2. Based on measurements, do KVL and KCL apply to the tested circuits (show work)?

Laboratory Manual for AC Electrical Circuit Analysis

3. In general, how would the phasor diagram of Figure 8.1 change if the frequency was raised?

4. In general, how would the phasor diagram of Figure 8.2 change if the frequency was lowered?

Passive Crossover

Objective

The frequency response of a simple two-way LC passive crossover is investigated. One output should attenuate signals above the crossover frequency while the other output should attenuate signals below the crossover frequency.

Theory Overview

In order to span the range of audible tones with accuracy, low distortion, and reasonable volume levels, loudspeaker systems are typically comprised of two or more transducers, each designed to cover only a portion of the frequency spectrum. As the spectrum is broken up into multiple segments, a circuit is needed to "steer" the proper signals to the appropriate transducers. Failure to do so may result in distortion or damage to the transducers. As inductors and capacitors exhibit a reactance that is a function of frequency, they are ideal candidates for this job. In this exercise, a simple two-way crossover is examined. It has one output for the high frequency transducer (tweeter) and for the low frequency transducer (woofer). In order to reduce the size of the components in this exercise, the impedance has been scaled upward by a factor of nearly 100. In place of the transducers, two resistive loads are used. This has the added advantage of not producing any sound in the lab! While real-world crossovers tend to be more complex than the one in this exercise, it will suffice to show the basic operation.

Equipment

(1) AC function generator	model:_________________	srn:_________________
(1) Oscilloscope	model:_________________	srn:_________________

Components

(1) 250 nF	actual:_________________
(1) 100 mH	actual:_________________
(2) 620 Ω	actual:_________________

     Laboratory Manual for AC Electrical Circuit Analysis

Schematics

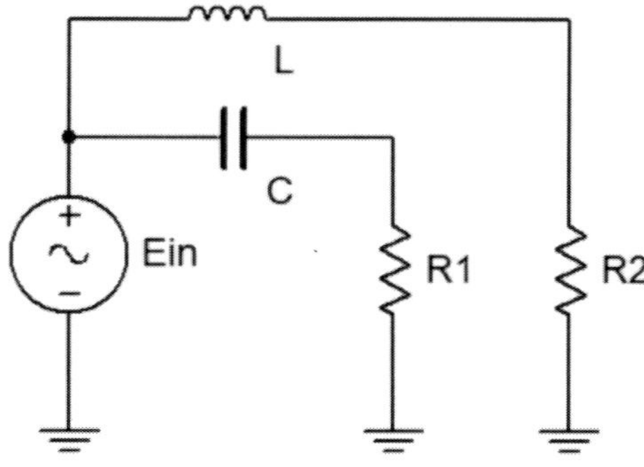

Figure 9.1

Procedure

1. The circuit of Figure 9.1 can be thought of as a pair of frequency dependent voltage dividers. X_L increases with frequency, thus attenuating high frequency signals reaching R2. Similarly, X_C increases with a decrease in frequency, thus attenuating low frequency signals reaching R1. (R2 takes the place of the woofer while R1 takes the place of the tweeter). The crossover frequency is the frequency where $R1=X_C$ and $R2=X_L$ (normally the same frequency for both). Using C=250 nF, L=100 mH, and R1=R2=620 Ω, determine the crossover frequencies and record in Table 9.1.

2. Using the voltage divider rule and *Ein*=2 V p-p, determine and record the theoretical voltage at output one (R1) for each frequency listed in Table 9.2. Be sure to include both magnitude and phase.

3. Build the circuit of figure 9.1 using R1=R2=620 Ω, L= 100 mH, and C=250 nF. Place one probe across the generator and another across output one (R1). Set the generator to a 2 V p-p sine wave at 50 Hz. Make sure that the *Bandwidth Limit* of the oscilloscope is engaged for both channels. This will reduce the signal noise and make for more accurate readings.

4. Measure and magnitude and phase shift of the output with respect to the input and record in Table 9.2. Repeat the measurements for the remaining frequencies in the table.

5. Repeat Steps two through four using the second output (R2) and Table 9.3.

6. On a single graph plot the magnitude response of both outputs with respect to frequency. On a separate graph plot the phase response of both outputs with respect to frequency.

Data Tables

$f_{tweeter}$	
f_{woofer}	

Table 9.1

Output One

Frequency	V1 Mag Theory	V1 θ Theory	V1 Mag Exp	V1 θ Exp
50				
70				
100				
200				
500				
1k				
2k				
5k				
10k				
15k				
20k				

Table 9.2

 Laboratory Manual for AC Electrical Circuit Analysis

Output Two

Frequency	V2 Mag Theory	V2 θ Theory	V2 Mag Exp	V2 θ Exp
50				
70				
100				
200				
500				
1k				
2k				
5k				
10k				
15k				
20k				

Table 9.3

Questions

1. Are the responses of the two outputs symmetrical? Do they need to be?

2. What is the maximum attenuation at the frequency extremes for the two outputs?

3. How might the attenuation be increased?

10

AC Superposition

Objective

This exercise examines the analysis of multi-source AC circuits using the superposition theorem. In particular, sources with differing frequencies will be used to illustrate the contributions of each source to the combined result.

Theory Overview

The superposition theorem can be used to analyze multi-source AC linear bilateral networks. Each source is considered in turn, with the remaining sources replaced by their internal impedance, and appropriate series-parallel analysis techniques employed. The resulting signals are then summed to produce the combined output signal. To see this process more clearly, the exercise will utilize two sources operating at different frequencies. Note that as each source has a different frequency, the inductor and capacitor appear as different reactances to the two sources.

Equipment

(2) AC function generators	model:_____________	srn:_______________
	model:_____________	srn:_______________
(1) Oscilloscope	model:_____________	srn:_______________

Components

(1) 100 nF   actual:_________________

(1) 10 mH   actual:_________________

(1) 1k Ω actual:_________________

(1) 50 Ω actual:_________________

Schematics

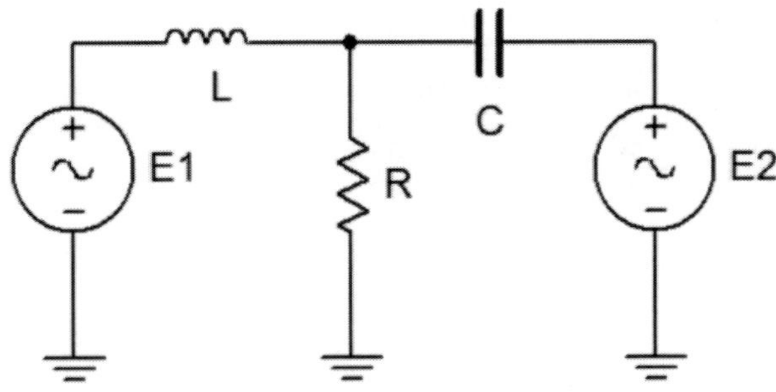

Figure 10.1

Procedure

1. Typical function generators have a 50 Ω internal impedance. These are not shown in the circuit of Figure 10.1. To test the superposition theorem, sources *E1* and *E2* will be examined separately and then together.

Source One Only

2. Consider the circuit of Figure 10.1 with C=100 nF, L=10 mH, R=1 kΩ, using only source *E1*=2 V p-p at 1 kHz and with source *E2* replaced by its internal impedance of 50 Ω. Using standard series-parallel techniques, calculate the voltages across *E1*, R, and *E2*. Remember to include the 50 Ω internal impedances in the calculations. Record the results in Table 10.1.

3. Build the circuit of Figure 10.1 using C=100 nF, L=10 mH, and R=1 kΩ. Replace *E2* with a 50 Ω resistor to represent its internal impedance. Set *E1* to 2 V p-p at 1 kHz, unloaded. Make sure that the *Bandwidth Limit* of the oscilloscope is engaged for both channels. This will reduce the signal noise and make for more accurate readings. Place probe one across *E1* and probe two across R. Measure the voltages across *E1* and R, and record in Table 10.1. Record a copy of the scope image. Move probe two across *E2* (the 50 Ω), measure and record this voltage in Table 10.1.

Source Two Only

4. Consider the circuit of Figure 10.1 using only source *E2*=2 V p-p at 10 kHz and with source *E1* replaced by its internal impedance of 50 Ω. Using standard series-parallel techniques, calculate the

voltages across *E1*, R, and *E2*. Remember to include the 50 Ω internal impedances in the calculations. Record the results in Table 10.2.

5. Replace the 50 Ω with source *E2* and set it to 2 V p-p at 10 kHz, unloaded. Replace *E1* with a 50 Ω resistor to represent its internal impedance. Place probe one across *E2* and probe two across R. Measure the voltages across *E2* and R, and record in Table 10.2. Record a copy of the scope image. Move probe two across *E1* (the 50 Ω), measure and record this voltage in Table 10.2.

Sources One and Two

6. Consider the circuit of Figure 10.1 using both sources, *E1*=2 V p-p at 1 kHz and *E2*=2 V p-p at 10 kHz. Add the calculated voltages across *E1*, R, and *E2* from Tables 10.1 and 10.2. Record the results in Table 10.3. Make a note of the expected maxima and minima of these waves and sketch how the combination should appear on the scope

7. Replace the 50 Ω with source *E1* and set it to 2 V p-p at 1 kHz, unloaded. **Both sources should now be active.** Place probe one across *E1* and probe two across R. Measure the voltages across *E1* and R, and record in Table 10.3. Record a copy of the scope image. Move probe two across *E2*, measure and record this voltage in Table 10.3.

Computer Simulation

8. Build the circuit of Figure 10.1 in a simulator. Using Transient Analysis, determine the voltage across the resistor and compare it to the theoretical and measured values recorded in Table 10.3. Be sure to include the 50 Ω source resistances in the simulation.

Data Tables

Source One Only

	Theory	Experimental	% Deviation
E_1			
E_2			
V_R			

Table 10.1

Source Two Only

	Theory	Experimental	% Deviation
E_1			
E_2			
V_R			

Table 10.2

Sources One and Two

	Theory	Experimental	% Deviation
E_1			
E_2			
V_R			

Table 10.3

Questions

1. Why must the sources be replaced with a 50 Ω resistor instead of being shorted?

2. Do the expected maxima and minima from step 6 match what is measured in step 7?

3. Does one source tend to dominate the 1 kΩ resistor voltage or do both sources contribute in nearly equal amounts? Will this always be the case?

11

AC Thevenin's Theorem

Objective

Thevenin's theorem will be examined for the AC case. The Thevenin source voltage and Thevenin impedance will be determined experimentally and compared to theory. Loads will be examined when driven by both an arbitrary circuit and that circuit's Thevenin equivalent to determine if the resulting load potentials are indeed identical. Both resistive and complex loads will be examined as well as well source impedances that are inductive or capacitive.

Theory Overview

Thevenin's theorem states that any linear single port (i.e., two terminals) network can be replaced by a single voltage source with series impedance. While the theorem is applicable to any number of voltage and current sources, this exercise will only examine single source circuits for the sake of simplicity. The Thevenin voltage is the open circuit output voltage. This may be determined experimentally by isolating the portion to be Thevenized and simply placing an oscilloscope at its output terminals. The Thevenin impedance is found by replacing all sources with their internal impedance and then applying appropriate series-parallel impedance simplification rules. If an impedance meter is available, an easy method of doing this in the lab is to replace the sources with appropriate impedance values and apply the impedance meter to the output terminals of the circuit portion under investigation.

Equipment

(1) AC function generator	model:____________	srn:____________
(1) Oscilloscope	model:____________	srn:____________
(1) Variable frequency impedance meter	model:____________	srn:____________
(1) Decade resistance box	model:____________	srn:____________

Components

(1) 100 nF	actual:____________
(1) 470 nF	actual:____________
(1) 10 mH	actual:____________
(1) 50 Ω	actual:____________

(1) 1.0 kΩ   actual:______________________

(1) 1.5 kΩ   actual:______________________

(1) 2.2 kΩ   actual:______________________

Schematics

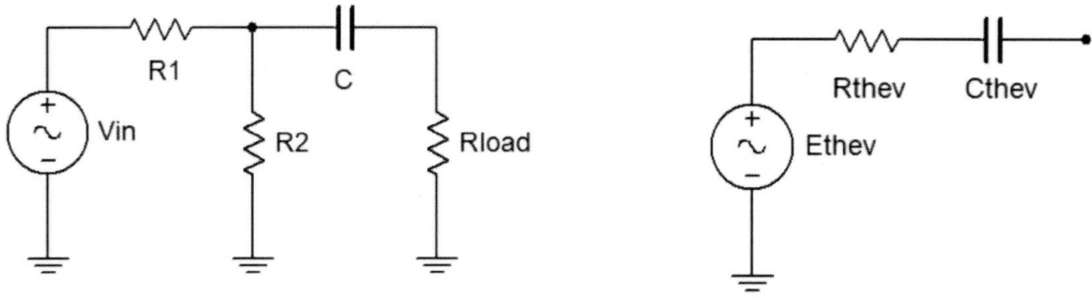

Figure 11.1

Figure 11.2

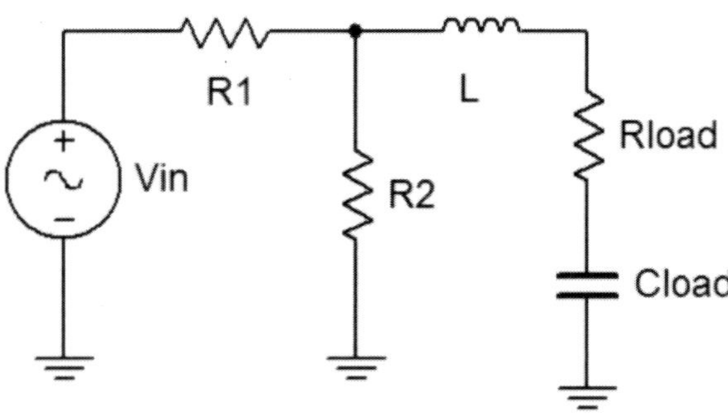

Figure 11.3

Procedure

1. For the circuit of figure 11.1, calculate the voltage across the 1 kΩ load using R1=1.5 kΩ, R2=2.2 kΩ, and C=470 nF, with a 2 V p-p 1 kHz source. Record this value in Table 11.1. Also calculate the expected Thevenin voltage and Thevenin impedance. Record these values in Table 11.2.

2. Build the circuit of figure 11.1 using R1=1.5 kΩ, R2=2.2 kΩ, Rload=1 kΩ and C=470 nF. Set the generator to a 1 kHz sine wave at 2 V p-p. Make sure that the *Bandwidth Limit* of the oscilloscope is engaged. This will reduce the signal noise and make for more accurate readings. Measure the load voltage and record in Table 11.1 as V_{Load} Original.

3. Remove the load and measure the unloaded output voltage. This is the experimental Thevenin voltage. Record it in Table 11.2.

4. Replace the voltage source with a 50 Ω resistor to represent its internal impedance. Set the impedance meter to 1 kHz and measure the resulting impedance at the open load terminals. This is the experimental Thevenin impedance. Record these values in Table 11.2 and compare with the theoretical values.

5. Using the decade resistance box and capacitor, build the Thevenin equivalent circuit of figure 11.2 and apply the 1 kΩ load resistor. Measure the load voltage and record in Table 11.1. Compare with the values of the original (non-Thevenized) circuit and determine the deviation between the original and Thevenized circuits.

6. To verify that Thevenin's theorem also works with an inductive source and a complex load, repeat steps 1 through 5 in like manner but using figure 11.3 with R1=1.5 kΩ, R2=2.2 kΩ, L=10 mH, Rload=1 kΩ with Cload=100 nF. Set the generator to a 10 kHz sine wave at 2 V p-p. Record results in Tables 11.3 and 11.4.

Data Tables

V_{load} Theory	
V_{load} Original	
V_{load} Thevenin	
% Deviation	

Table 11.1

	Theory	Experimental	% Deviation
$E_{Thevenin}$			
$Z_{Thevenin}$			

Table 11.2

                                    Laboratory Manual for AC Electrical Circuit Analysis

V_{load} Theory	
V_{load} Original	
V_{load} Thevenin	
% Deviation	

Table 11.3

	Theory	Experimental	% Deviation
$E_{Thevenin}$			
$Z_{Thevenin}$			

Table 11.4

Questions

1. How does the AC version of Thevenin's theorem compare with the DC version?

2. Would the Thevenin equivalent circuits be altered if the source frequency was changed? If so, why?

3. Based on the results of this exercise, would you expect Norton's theorem for AC to behave similarly to its DC case?

12

AC Maximum Power Transfer

Objective

In this exercise, maximum power transfer to the load will be examined for the AC case. Both the load's resistive and reactive components will be independently varied to discover their effect of load power and determine the values required for maximum load power.

Theory Overview

In the DC case, maximum power transfer is achieved by setting the load resistance equal to the source's internal resistance. This is not true for the AC case. Instead, the load should be set to complex conjugate of the source impedance, the complex conjugate having the same magnitude as the original but with the opposite sign for the angle. By using the complex conjugate, the load and source reactive components will cancel out leaving a purely resistive circuit similar to the DC case. When calculating the true load power, care must be taken to remember that the load voltage appears across a complex load impedance. Only the real portion of this voltage appears across the resistive component, and only the resistive component dissipates power.

Equipment

(1) AC function generator	model:____________	srn:____________
(1) Oscilloscope	model:____________	srn:____________
(1) Decade resistance box	model:____________	srn:____________
(1) Impedance meter	model:____________	srn:____________

Components

(1) 10 mH   actual:____________

(1) 1k Ω actual:____________

(1) 50 Ω actual:____________

(1) 100 nF   actual:____________

(1) 47 nF   actual:____________

(1) 33 nF   actual:____________

(1) 22 nF   actual:____________

(1) 10 nF   actual:____________   Assorted capacitors in the 1 nF region

Schematics

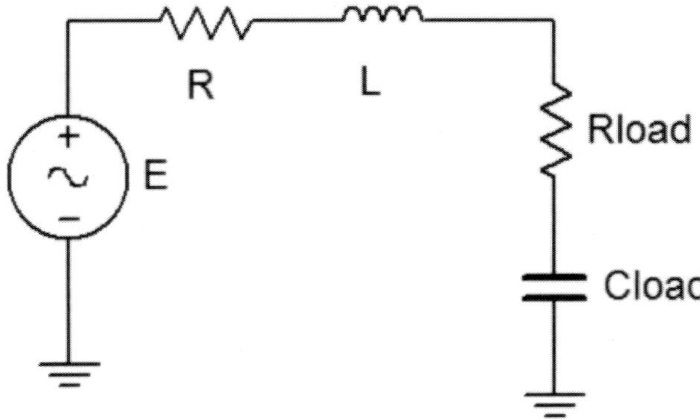

Figure 12.1

Procedure

1. Build the circuit of figure 12.1 using R=1 kΩ and L= 10 mH, but leaving off the load components. Replace the generator with a 50 Ω resistor and determine the effective source impedance at 10 kHz using the impedance meter. Record this value in Table 12.1, including both magnitude and phase. Determine the load impedance which should achieve maximum power transfer according to the theorem and record in Table 12.1. Finally, determine values for R_{load} and C_{load} to achieve this load impedance and record in Table 12.1, also copying the resistance value to the first R_{load} entry of Table 12.2.

Testing R_{load}

2. Replace the 50 Ω resistor with the generator. Insert the decade resistance box in the position of R_{load} and set it to the value calculated in Table 12.1. For C_{load}, use the value calculated in Table 12.1. Use multiple capacitors if necessary to achieve a close value. Set the generator to 10 volts peak at 10 kHz, making sure that the amplitude is measured on the oscilloscope with the generator loaded by the circuit. Make sure that the *Bandwidth Limit* of the oscilloscope is engaged for the channel. This will reduce the signal noise and make for more accurate readings.

3. Measure the magnitude of the load voltage and record in Table 12.2. Also compute the expected load voltage from theory and the load power based on the measured load voltage and record in Table 12.2. Repeat these measurements and calculations for the remaining load resistance values in the table.

4. Return the decade box to the value calculated in Table 12.1. For C_{load}, insert the first capacitor listed in Table 12.3. Repeat step four for each load capacitance in Table 12.3, calculating and recording the required results using Table 12.3.

5. Generate a plot of P_{load} with respect to R_{load} and another of P_{load} with respect to C_{load}.

Data Tables

Z_{source}	
Z_{load}	
R_{load}	
C_{load}	

Table 12.1

Variable R_{load}

R_{load}	V_{load} Theory	V_{load} Exp	P_{load} Exp
100			
400			
600			
800			
1.2 k			
1.8 k			
3 k			
10 k			

Table 12.2

Variable C_{load}

C_{load}	V_{load} Theory	V_{load} Exp	P_{load} Exp
1 nF			
3.3 nF			
10 nF			
33 nF			
47 nF			
100 nF			

Table 12.3

Questions

1. In general, given a certain source impedance, what load impedance will achieve maximum load power?

2. Will achieving maximum load power also achieve maximum efficiency? Explain.

3. If the experiment was repeated using a frequency of 5 kHz, how would the graphs change, if at all?

13

Series Resonance

Objective

This exercise investigates the voltage relationships in a series resonant circuit. Of primary importance are the establishment of the resonant frequency and the quality factor, or Q, of the circuit with relation to the values of the R, L, and C components.

Theory Overview

A series resonant circuit consists of a resistor, a capacitor, and an inductor in a simple loop. At some frequency the capacitive and inductive reactances will be of the same magnitude, and as they are 180 degrees in opposition, they effectively nullify each other. This leaves the circuit purely resistive, the source "seeing" only the resistive element. Consequently, the current will be at a maximum at the resonant frequency. At any higher or lower frequency, a net reactance (the difference between X_L and X_C) must be added to the resistor value, producing a higher impedance and thus, a lower current. As this is a simple series loop, the resistor's voltage will be proportional to the current. Consequently, the resistor voltage should be a maximum at the resonant frequency and decrease as the frequency is either increased or decreased. At resonance, the resistor value sets the maximal current and consequently has a major effect on the voltages developed across the capacitor and inductor as well as the "tightness" of the voltage versus frequency curve: The smaller the resistance, the tighter the curve and the higher the voltage seen across the capacitor and inductor. The Q of the circuit can be defined as the ratio of the resonant reactance to the circuit resistance, $Q=X/R$, which also corresponds to the ratio of the resonant frequency to the circuit bandwidth, $Q=F_0/BW$.

Equipment

(1) AC function generator	model:___________________	srn:___________________
(1) Oscilloscope	model:___________________	srn:___________________

Components

(1) 10 nF	actual:___________________
(1) 10 mH	actual:___________________

(1) 47 Ω   actual:______________________

(1) 470 Ω   actual:______________________

Schematics

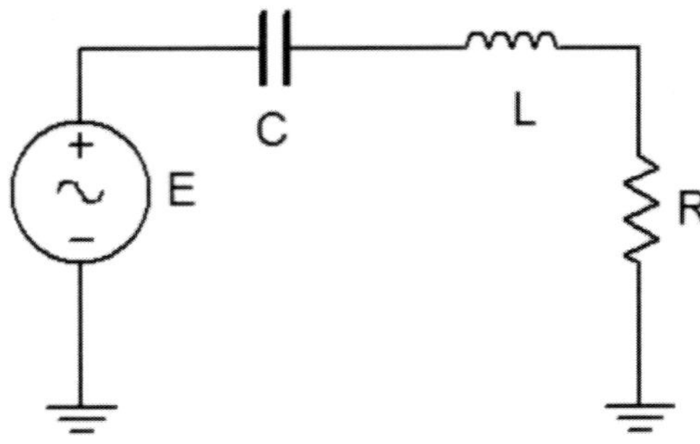

Figure 13.1

Procedure

Low Q Circuit

1. Using Figure 13.1 with R=470 Ω, L= 10 mH, and C=10 nF, determine the theoretical resonance frequency and Q, and record the results in Table 13.1. Based on these values determine the upper and lower frequencies defining the bandwidth, f_1 and f_2, and record them in Table 13.1.

2. Build the circuit of Figure 13.1 using R=470 Ω, L=10 mH and C=10 nF. Place a probe across the resistor. Set the output of the generator to a 1 V p-p sine wave. Set the frequency to the theoretical resonance frequency of Table 13.1. Make sure that the *Bandwidth Limit* of the oscilloscope is engaged for both channels. This will reduce the signal noise and make for more accurate readings.

3. Adjust the frequency in small amounts, up and down, until the maximum voltage is found. This is the experimental resonance frequency. Record it in Table 13.1. Note the amplitude (it should be approximately equal to the source voltage of 1 V p-p). Sweep the frequency above and below the resonance frequency until the experimental f_1 and f_2 are found. These will occur at a voltage amplitude of approximately 0.707 times the resonant voltage (i.e., the half-power points). Record these frequencies in Table 13.1. Also, determine and record the experimental Q based on the experimental f_0, f_1, and f_2.

4. Transcribe the experimental frequencies of Table 13.1 to the top three entries of Table 13.2. For all of the frequencies in Table 13.2, measure and record the voltage across the resistor. Also measure and record the inductor and capacitor voltages. Note that the inductor and capacitor will have to be swapped with the resistor position in order to maintain proper ground reference with the oscilloscope.

5. Based on the data from Table 13.2, plot V_R, V_C, and V_L as a function of frequency.

6. Change R to 47Ω and repeat steps 1 through 5 but using Tables 13.3 and 13.4 for high Q.

Computer Simulation

7. Build the circuit of Figure 13.1 in a simulator. Using AC Analysis, plot the voltage across the resistor from 1 kHz to 100 kHz for both the high and low Q cases and compare them to the plots derived from Tables 13.2 and 13.4. Be sure to include the 50 Ω source resistance and coil resistance in the simulation.

Data Tables

Low Q Circuit

	Theory	Experimental	% Deviation
f_o			
Q			
f_1			
f_2			

Table 13.1

Frequency	V_R	V_C	V_L
$f_0=$			
$f_1=$			
$f_2=$			
1 kHz			
5 kHz			
8 kHz			
12 kHz			
20 kHz			
30 kHz			
50 kHz			
100 kHz			

Table 13.2

High Q Circuit

	Theory	Experimental	% Deviation
f_o			
Q			
f_1			
f_2			

Table 13.3

Frequency	V_R	V_C	V_L
$f_0=$			
$f_1=$			
$f_2=$			
1 kHz			
5 kHz			
8 kHz			
12 kHz			
20 kHz			
30 kHz			
50 kHz			
100 kHz			

Table 13.4

Questions

1. What is the effect of changing resistance on Q?

2. Are the V_C and V_L curves the same as the V_R curves? If not, why?

3. In practical terms, what sets the limit on how high Q may be?

14

Parallel Resonance

Objective

This exercise investigates the voltage relationships in a parallel resonant circuit. Of primary importance are the establishment of the resonant frequency and the quality factor, or Q, of the circuit with relation to the values of the R, L, and C components.

Theory Overview

A parallel resonant circuit consists of a resistor, a capacitor, and an inductor in parallel, typically driven by a current source. At some frequency the capacitive and inductive reactances will be of the same magnitude, and as they are 180 degrees in opposition, they effectively nullify each other. This leaves the circuit purely resistive, the source "seeing" only the resistive element. At any lower or higher frequency the inductive or capacitive reactance will shunt the resistance. The result is a maximum impedance magnitude at resonance, and thus, a maximum voltage. Any resistance value in series (such as the inductor's coil resistance) should be transformed into a parallel resistance in order to gauge its effect on the system voltage. The combined parallel resistance sets the Q of the circuit and can be defined as the ratio of the combined resistance to the resonant reactance, $Q=R/X$, which also corresponds to the ratio of the resonant frequency to the circuit bandwidth, $Q=f_0/BW$.

Equipment

(1) AC function generator	model:____________	srn:____________
(1) Oscilloscope	model:____________	srn:____________

Components

(1) 10 nF	actual:____________
(1) 10 mH	actual:____________
(1) 2.2 kΩ	actual:____________
(1) 10 kΩ	actual:____________
(1) 100 kΩ	actual:____________

Schematics

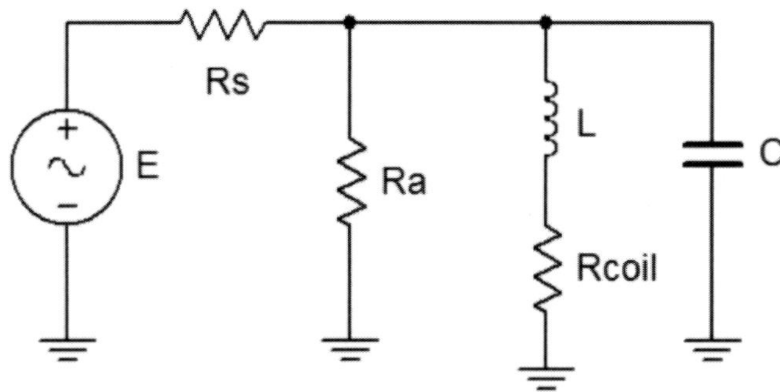

Figure 14.1

Procedure

1. Using Figure 14.1 with Rs=100 kΩ, Ra=2.2 kΩ, L=10 mH, Rcoil=7 Ω and C=10 nF, determine the theoretical resonance frequency and Q, and record the results in Table 14.1. Based on these values determine the upper and lower frequencies defining the bandwidth, f_1 and f_2, and record them in Table 14.1 also.

2. Build the circuit of Figure 14.1 using Rs=100 kΩ, Ra=2.2 kΩ, L=10 mH and C=10 nF. Set the output of the generator to a 10 V p-p sine wave at the theoretical resonant frequency. The large value of Rs associated with the voltage source will make it appear as a current source equal to approximately 100 μA p-p, assuming the parallel branch impedance is much less than Rs. Place a probe across the parallel branch. Set the frequency to the theoretical resonance frequency of Table 14.1. Make sure that the *Bandwidth Limit* of the oscilloscope is engaged for both channels. This will reduce the signal noise and make for more accurate readings.

3. Adjust the frequency in small amounts, up and down, until the maximum voltage is found. This is the experimental resonant frequency. Record it in Table 14.1. Note the amplitude. Sweep the frequency above and below the resonance frequency until the experimental f_1 and f_2 are found. These will occur at a voltage amplitude of approximately 0.707 times the resonant voltage (i.e., the half-power points). Record these frequencies in Table 14.1. Also, determine and record the experimental Q based on the experimental f_0, f_1, and f_2.

4. Transcribe the experimental frequencies of Table 14.1 to the top three entries of Table 14.2. For all of the frequencies in Table 14.2, measure and record the voltage across the parallel branch.

5. Based on the data from Table 14.2, plot the parallel branch voltage as a function of frequency.

6. For high Q, change Ra to 10 kΩ and repeat steps 1 through 5 but using Tables 14.3 and 14.4.

Data Tables

Low Q Circuit

	Theory	Experimental	% Deviation
f_0			
Q			
f_1			
f_2			

Table 14.1

Frequency	$V_{Parallel}$
$f_0=$	
$f_1=$	
$f_2=$	
1 kHz	
5 kHz	
8 kHz	
12 kHz	
20 kHz	
30 kHz	
50 kHz	
100 kHz	

Table 14.2

                    Laboratory Manual for AC Electrical Circuit Analysis

High Q Circuit

	Theory	Experimental	% Deviation
f_0			
Q			
f_1			
f_2			

Table 14.3

Frequency	$V_{Parallel}$
$f_0=$	
$f_1=$	
$f_2=$	
1 kHz	
5 kHz	
8 kHz	
12 kHz	
20 kHz	
30 kHz	
50 kHz	
100 kHz	

Table 14.4

Questions

1. What is the effect of changing resistance on Q?

2. Are f_1 and f_2 spaced symmetrically around f_0?

3. In practical terms, what sets the limit on how high Q may be?

15

A Loudspeaker Impedance Model

Objective

This exercise investigates the impedance magnitude and phase of typical dynamic moving voice coil loudspeakers. The impedance is dominated by the voice coil resistance at DC, by the mechanical resonant system which produces a parallel resonant equivalent in the bass region, and finally by the voice coil inductance at the highest frequencies.

Theory Overview

Loudspeakers are typically specified with a nominal impedance of four or eight ohms, although other values are possible. The actual impedance of a typical loudspeaker can vary widely from this rating. This exercise examines the impedance of two different permanent magnet-voice coil type transducers with respect to frequency. Both amplitude and phase are important to consider. Devices of this type normally exhibit a resonant peak in the bass end and a gradual rise in magnitude as frequency increases. The phase angle is at times capacitive, inductive, and also resistive. Typical loudspeaker impedance is not nearly as consistent as simple resistors. The resulting complex impedance can present a much more challenging load for an audio amplifier than a simple ideal eight ohm resistance.

Equipment

(1) AC function generator	model:________________	srn:________________
(1) Oscilloscope	model:________________	srn:________________
(1) DMM	model:________________	srn:________________

Components

(1) 1 kΩ   actual:________________

(1) 6" or larger woofer

(1) 4" general purpose loudspeaker

Schematic

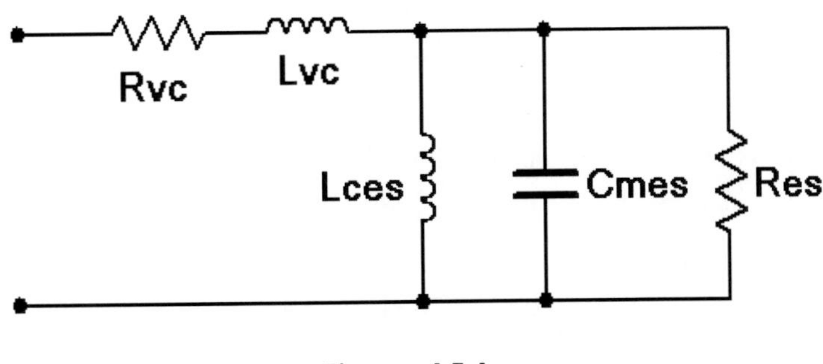

Figure 15.1

Procedure

1. Figure 15.1 shows a typical model of a dynamic moving voice coil loudspeaker mounted on an infinite baffle[1]. Rvc and Lvc represent the resistance and inductance of the of the voice coil. Lces represents the driver compliance (i.e., springiness of suspension and air), Cmes represents the driver's mass, and Res represents the frictional losses of the suspension. At DC, the impedance will equal Rvc. At high frequencies Lvc will dominate the response. The parallel resonant portion will create an impedance peak in the bass. This is normally referred to as the free air resonance of the driver, or f_s.

2. Using the woofer, measure the DC resistance using the DMM and record the value in Table 15.1.

3. In order to measure both the magnitude and phase of the impedance across frequency, it is desirable to drive the loudspeaker with a fixed current source. By measuring the voltage across the loudspeaker with an oscilloscope, both amplitude and time delay can be measured, and thus both magnitude and phase of impedance can be calculated. A current source may be approximated by placing a large resistor in series with the function generator. If the resistance value is many times greater than the loudspeaker impedance, the loudspeaker may be ignored to a first approximation. Therefore, virtually all of the generator voltage drops across the series resistor. For this exercise, a 1 kΩ value will suffice. For all measurements, simply place a 1 kΩ resistor in series with the generator and the loudspeaker under test. Place oscilloscope probes at both ends of the resistor (i.e., input signal and loudspeaker signal). Make sure that the *Bandwidth Limit* of the oscilloscope is engaged for both channels. This will reduce the signal noise and make for more accurate readings.

4. Hook up the woofer between the resistor and ground. Make sure that the woofer is magnet-side down, with the cone facing up, and unobstructed.

---

[1] Adapted from R.H.Small, Direct-Radiator Loudspeaker System Analysis, Journal of the Audio Engineering Society, June 1972

5. First, find the resonant frequency. To do this, set the output of the generator to approximately 100 Hz, sine wave, and 10 volts peak. Adjust the frequency in small amounts, up and down, until the maximum voltage is found. This is the experimental free air resonant frequency. Record it in Table 15.1. Note the amplitude. Sweep the frequency above and below the resonant frequency until the experimental f_1 and f_2 are found. These will occur at a voltage amplitude of approximately 0.707 times the resonant voltage (i.e., the half-power points). Record these frequencies in Table 15.2. Copy the three frequencies to the first three entries of Table 15.3.

6. For the frequencies in Table 15.3 determine the amplitude and time delay at the loudspeaker and compute the phase shift. Be sure to include whether the loudspeaker signal is leading (+) or lagging (-) the source.

7. Swap the woofer with the general purpose loudspeaker and repeat steps 2 through 6 using Tables 15.4 through 15.6. Plot the magnitude and phase of each device on semi-log paper. Also, try to estimate values for the schematic model of Figure 15.1.

Data Tables

R_{VC}	

Table 15.1

	Frequency	Voltage
f_s		
f_1		
f_2		

Table 15.2

Frequency	Amplitude	Delay Time	Phase Shift
$f_s=$			
$f_1=$			
$f_2=$			
20 Hz			
25 Hz			
30 Hz			
40 Hz			
50 Hz			
60 Hz			
75 Hz			
100 Hz			
200 Hz			
500 Hz			
1 kHz			

Table 15.3

Rvc	

Table 15.4

	Frequency	Voltage
f_s		
f_1		
f_2		

Table 15.5

Frequency	Amplitude	Delay Time	Phase Shift
$f_s=$			
80 Hz			
120 Hz			
150 Hz			
170 Hz			
200 Hz			
500 Hz			
1 kHz			
3 kHz			

Table 15.6

Questions

1. Is the resulting impedance always resistive?

2. Is the resulting impedance always inductive?

3. How do the two curves compare?

Appendix A: Plotting Phasors with a Spreadsheet

While it can be instructive to create phasor plots the old-fashioned way with graph paper, ruler and protractor, the results are seldom as nice as those produced by computer plotting programs. Dedicated scientific plotting software may be the most flexible route (check out SciDAVis) but the following method produces workable results with almost any commonly available spreadsheet. The example below uses Open Office 4 but the process will be similar using other tools such as Excel. The precise menu and dialog options will vary from program to program and from version to version.

1. First, prepare your data. Typically, you'll want to plot a collection of voltages or currents that are in polar form (magnitude ∟ angle°). For this technique, the data need to be converted to rectangular form (real $\pm$ j imaginary). While it's possible to have the spreadsheet do the conversion, for simplicity sake and the fact that it's a trivial operation for a scientific/engineering calculator, we'll just use rectangular form data entry in the spreadsheet. For example, let's say you have measured three voltages: 1∟135° (second quadrant), 1∟45° (first quadrant) and 1∟-45° (fourth quadrant). These would be converted to: (-0.707 + j 0.707), (0.707 + j 0.707) and (0.707 - j 0.707). The real portions correspond to the horizontal X axis and the imaginary portions correspond to the vertical Y axis.

2. A scatter plot will be used to create the phasor diagram. By default, a scatter plot will draw a line from each coordinate pair to the next, moving sequentially through the data. If we were to enter these three pairs of data as is, the scatter plot would draw a line from point to point to point, not at all what we want. What we want is a set of lines radiating individually from the origin (0,0) to each of the three points. We can get the scatter plot to do this by simply inserting a (0,0) data pair between each item. In reality, the plot line is retracing the vector but we won't notice that. Also, it is imperative that no manner of curve fit be used, rather, simple first-order straight line segments should be used to "connect the dots".

3. Open a new worksheet. In general, column A will contain the real portions and column B will contain the imaginary portions. In the first row (row 1), enter the text for the legend. Starting in the second row (row 2), enter the origin data for columns A and B, i.e., enter 0 for these cells. In the next row, enter the real and imaginary values for the first vector. In like fashion, enter origin and vectors values on subsequent rows. For the example above, the columns would look like:

0	0
-0.707	0.707
0	0
0.707	0.707
0	0
0.707	-0.707

4. Select/highlight all of the data (click the first cell, in the upper left corner, and drag the mouse over all of the cells used).

5. Select the Insert menu and choose Chart. Choose **XY Scatter** chart from the various types available. Select points with simple straight lines (no smoothing), set the titles and so forth, and finish.

6. You can customize the appearance of the chart. In general, you can edit items by simply double-clicking on the item or by using a right-mouse click to bring up a property menu. This will allow you to add or alter gridlines, axes, etc.

7. Once your chart is completed, you may wish to save the worksheet for future reference. To insert the chart into a lab report, select the chart by clicking on it, copy it to the clipboard (Ctrl+C), select the insertion point in the lab report, and paste (Ctrl+V).

8. **This is very important:** The spreadsheet will scale and adjust the axes to optimize the view. It will not necessarily keep the XY scaling consistent (i.e., the same volts/centimeter for each axis). Ignoring this fact may yield a plot that appears squashed and lines that should be perpendicular won't appear as such. To correct for this, you must adjust the axis ranges or the aspect ratio of the chart so that the horizontal and vertical scales are equally sized. This is easiest to do if both X and Y axis grid lines are selected and set to the same values (scaling will be correct when the grid appears square).

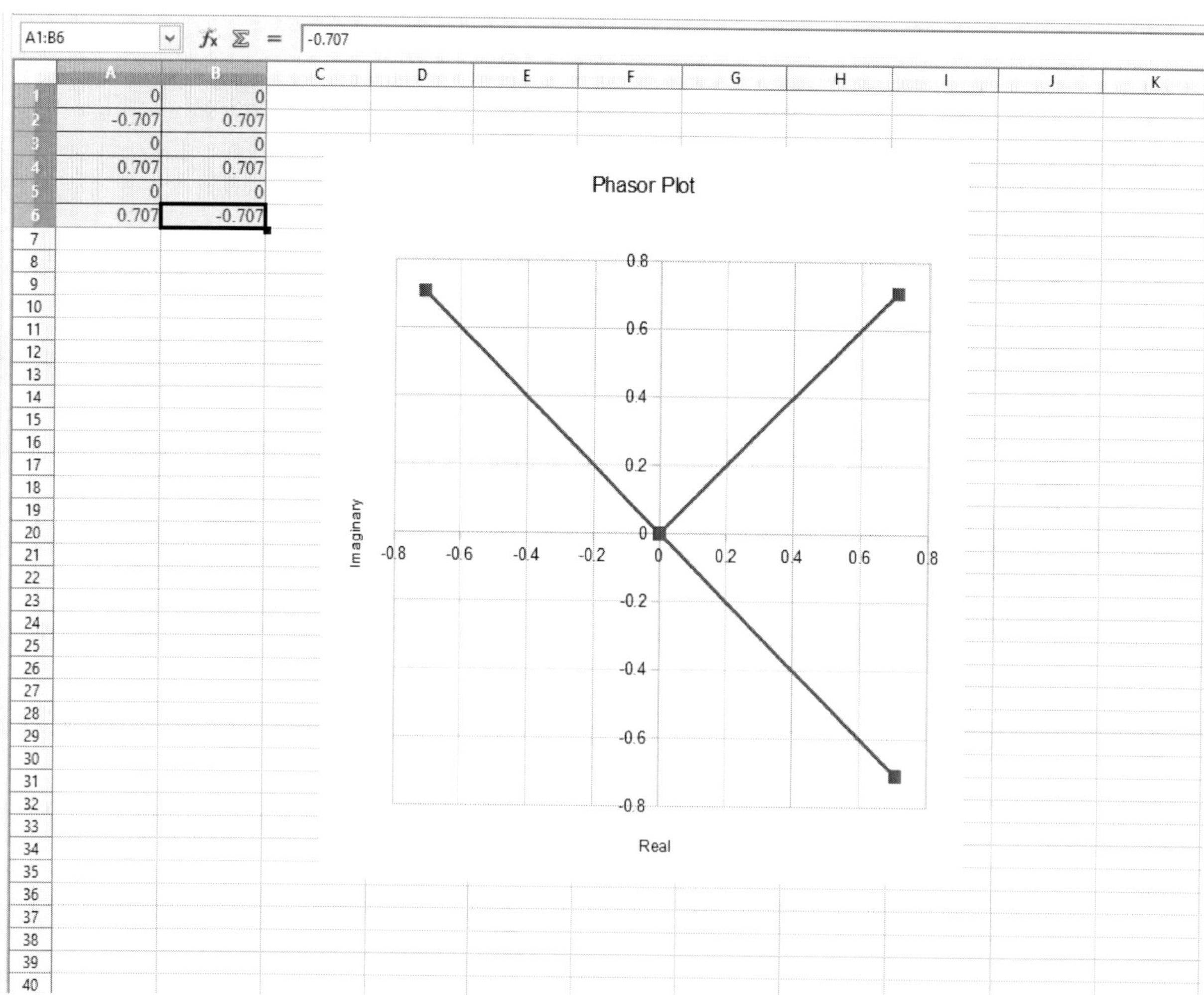

Laboratory Manual for AC Electrical Circuit Analysis

Made in United States
North Haven, CT
12 January 2022

14655984R00054